CAMBRIDGE LIBRARY COLLECTION

Books of enduring scholarly value

Technology

The focus of this series is engineering, broadly construed. It covers technological innovation from a range of periods and cultures, but centres on the technological achievements of the industrial era in the West, particularly in the nineteenth century, as understood by their contemporaries. Infrastructure is one major focus, covering the building of railways and canals, bridges and tunnels, land drainage, the laying of submarine cables, and the construction of docks and lighthouses. Other key topics include developments in industrial and manufacturing fields such as mining technology, the production of iron and steel, the use of steam power, and chemical processes such as photography and textile dyes.

A Description of Westminster Bridge

The construction of the first Westminster Bridge, upon which Wordsworth composed his famous sonnet, presented many challenges in terms of the materials and methods with which a sturdy bridge could be built in tidal water and on a gravelly riverbed. A number of candidates presented their surveys to the commissioners of the bridge, but it was the Swiss-born Charles Labelye (1705–62) who was appointed to oversee construction in 1738. The bridge opened to traffic in 1750. This 1751 publication expands upon the shorter work that Labelye had prepared in 1739 to address the laying of the foundations. Significantly, he used caissons – vast wooden structures sunk into the riverbed – within which the stone piers were built. Although the promised illustrations did not appear in this work, the book provides a valuable insight into the technical problems of a major engineering project, and the solutions available at that time.

Cambridge University Press has long been a pioneer in the reissuing of out-of-print titles from its own backlist, producing digital reprints of books that are still sought after by scholars and students but could not be reprinted economically using traditional technology. The Cambridge Library Collection extends this activity to a wider range of books which are still of importance to researchers and professionals, either for the source material they contain, or as landmarks in the history of their academic discipline.

Drawing from the world-renowned collections in the Cambridge University Library and other partner libraries, and guided by the advice of experts in each subject area, Cambridge University Press is using state-of-the-art scanning machines in its own Printing House to capture the content of each book selected for inclusion. The files are processed to give a consistently clear, crisp image, and the books finished to the high quality standard for which the Press is recognised around the world. The latest print-on-demand technology ensures that the books will remain available indefinitely, and that orders for single or multiple copies can quickly be supplied.

The Cambridge Library Collection brings back to life books of enduring scholarly value (including out-of-copyright works originally issued by other publishers) across a wide range of disciplines in the humanities and social sciences and in science and technology.

A Description of Westminster Bridge

To Which are Added,
an Account of the Methods Made Use of
in Laying the Foundations of its Piers
and an Answer to the Chief Objections,
that Have Been Made Hitherto.

CHARLES LABELYE

CAMBRIDGE
UNIVERSITY PRESS

University Printing House, Cambridge, CB2 8BS, United Kingdom

Cambridge University Press is part of the University of Cambridge.
It furthers the University's mission by disseminating knowledge in the pursuit of education, learning and research at the highest international levels of excellence.

www.cambridge.org
Information on this title: www.cambridge.org/9781108071956

This edition first published 1751
This digitally printed version 2014

ISBN 978-1-108-07195-6 Paperback

A

DESCRIPTION

OF

Weſtminſter Bridge.

To which are added,

An Account of the Methods made uſe of in laying the Foundations of its Piers.

AND

An Anſwer to the chief Objections, that have been made thereto.

WITH

An APPENDIX,

CONTAINING

Several Particulars, relating to the ſaid BRIDGE, or to the Hiſtory of the Building thereof.

AS ALSO

Its Geometrical Plans, and the Elevation of one of the Fronts, as it is finiſhed, Correctly engraven on two large COPPER-PLATES.

—— *Quod optanti divum promittere nemo*
Auderet, volvenda Dies en attulit ultro. VIRG.

By *CHARLES LABELYE.*

LONDON:
Printed by W. STRAHAN, for the AUTHOR.
MDCCLI.

PREFACE.

OON after the Right Honourable, &c. the Commiſſioners, appointed by the Authority of Parliament, for building and maintaining *Weſtminſter Bridge*, had fixed upon a Deſign, and appointed ſuch Officers and Artificers as they thought proper, for the Direction and Execution of the Works; ſeveral of the Commiſſioners (and among them ſome well known to be good Judges in Civil and Military Architecture in general, and Bridge building in particular) deſired I would draw up an Account of the Method I intended to follow in laying the Foundations of the Piers of *Weſtminſter Bridge*, and give the Reaſons why that Method (which I had the Honour of explaining

plaining twice, upon Models before the Board) was preferable to other Methods proposed and made use of in the Building of Bridges.

On the 11th of *May*, 1738, I received an express Order of that Honourable Board for that Purpose. In Obedience to their Commands, I began soon after to write a short Account of this Method, but the Works requiring my constant Attendance, I was not able to compleat and publish it till *May*, 1739. Besides, finding that several People made Objections (even to the Possibility of putting in Practice those very Methods) I thought it best to postpone that Account, till after the first Pier should be compleated, by the Execution of which, most of those Objections would be destroyed; being almost sure that the most plausible Arguments, which could be offered before that time, would hardly prevail over some favourite Notions.

The short Account which I then published, was so well received by the Publick, that the whole Edition was soon disposed of, and ever since I have had frequent Requests to republish it, or a fuller Account; which for several Reasons (and among others for want of Leisure) I thought proper to defer till the Bridge was intirely finished, and opened to the Publick. Since which, the Board of Commissioners was pleased on *February*. the 26th last past, to give me a new Order for the drawing up, Printing, and Publishing the following Description and

Ac-

Accouns. Besides a short Description of *Westminster Bridge* (of which I have annex'd a correct Plan and Elevaion) the Reader will find in the following Sheets, proper mention made at first of the several Facts which induced me to make use of the Methods which I have followed, with all the Success imaginable, in the Building the Piers of *Westminster Bridge*, of which Method I have annexed as plain an Account as the Nature of such things will allow without Schemes or Figures, reserving a fuller Account to a larger Work, the Contents whereof I have given at the End. I have also inserted several Particulars relating to the Bridge, and some Accounts to give the Reader an Idea of the Quantity of Materials employed therein. Some of those Accounts have indeed been printed before, but having been most wretchedly mangled, and made absolute nonsense, by the Compilers of the Magazines, and other Phamphlets, I thought it was proper to repeat them here, correctly express'd.

I am very sensible that I should make some Apology for speaking of myself so much as I am obliged to do in this Relation, if I did not do it in Obedience to the Orders of the Commissioners, which direct me to give an Account of such Things as I have either done myself, or directed to be done.

As I will not be drawn into a Paper War upon this or any other Account whatsoever ; I shall never lose any Time in replying to any

thing

thing that may appear hereafter by way of Remarks upon, or of Anſwer to the Whole, or any Part of this Deſcription, and Accounts.

Laſtly, I hope the Readers will make me large Allowance as to my Stile and Diction, when they know I was neither born nor bred in *England*, but in *Swiſſerland* ; and never heard a Word of *Engliſh* ſpoken, till I was near twenty Years of Age.

ERRATA.

Page	Line	inſtead of	Read
9	19	not exceed	not much exceed
16	19	the Piles which were	where the Piles were
28	30	Frame	Frames
55	32	Same	Some
58	7	where what	where
58	17	three	two
67	7	Commiſſioner	Commiſſioners
71	1	proer	proper
75	15	all the Piers	all the 14 Piers
82	21	*Straum*	*Stratum*
96	23	not above	not much above
97	21	of Materials	of the ſeveral Materials

A

DESCRIPTION

OF

Westminster Bridge.

BY an Act of Parliament past in the 11th Year of his present Majesty's Reign, it was enacted, that the (then intended) Bridge should be built from the *Woolstaple*, or thereabouts, in the Parish of St. *Margaret*, in *Westminster*, to the opposite Shore, in the County of *Surry*.

In Consequence of this Determination, and in Obedience to the Orders of the Commissioners, several Observations were made to find out the Direction of the Stream in that Part of the River, at different Times of the Flood and Ebb; and accurate Surveys of the adjacent

Shores,

Shores, were taken by myſelf and others, which having thoroughly examined, I laid before the Board what Situation appeared the moſt advantageous; which being approved, proper Marks were made on each Shore, in order to place the Bridge in that Situation, which is as follows:

The Poſition of the new Bridge, in reſpect to the Points of the Compaſs, is almoſt due Eaſt and Weſt. And the Streams of the Tide (both Flood and Ebb) do paſs thro' the Arches, without any ſenſible Obliquity. This Poſition of the Bridge has alſo this great Advantage, that if a Street be ever made through the City of *Weſtminſter*, in the Continuation of the Direction of the Bridge, that Street will open a View to, or terminate in St. *James*'s-*Park*, at the End of *Princes-Court*, near *Story*'s-*Paſſage*.

The Breadth of the River at high Water, at the Place where the Bridge is built, was meaſured two different ways, firſt, by Trigonometry, *viz.* having meaſured a Baſis of near 700 Feet along the *Surry* Shore, and obſerved the proper Angles, the Calculation gave for the Diſtance intercepted between the *Woolſtaple Dock*, and the oppoſite Wharf, 1216 Feet. 2*dly*. It was actually meaſured with a wooden Beam of 100 Feet in Length, laid on the Surface of the River, at dead low Water, and it was found 1223 Feet; whence taking a mean between the Trigonometrical, and this actual Menſuration, which is likely to exceed the Truth

Truth (on Account of the Difficulty of laying the Beam preciſely in the ſame Line) it may ſafely be concluded, that the Breadth of the River at that Place is nearly 1220 Feet; which is about 300 Feet wider than at *London Bridge*, very near 400 Feet wider than at the *Horſe Ferry*, and about 200 Feet narrower than over-againſt *Whitehall* and *Scotland Yard*, on Account of the elbowing of the River, over-againſt thoſe laſt mentioned Places.

The Stream of the River, where the Bridge is built, is divided into two Channels; in the one, called the Ebb Channel, which runs along near the *Surry* Shore, the Water is about 8 Feet in Depth at a common low Water; and in the other, call'd the Flood Channel, which runs along near *Weſtminſter* Shore, the Water is about 6 Feet deep at a common low Water. Theſe two Channels are divided by a Shoal of a conſiderable Length and Breadth, made up of Sand, and of the Waſhing or Silting of the River. Upon the Shoal, the Water is ſeldom more than 4 Feet in Depth, at a common Low Water.

The Riſe of the Tide in this Part of the River, is very uncertain, being ſeldom leſs than 7 Feet at Neap Tides, and droughty Seaſons, eſpecially when the Wind is upon any Point of the Compaſs, between the South and the Weſt; and the Riſe of the Tide is ſeldom more than 17 Feet at extraordinary Spring Tides, joined with the Land Waters, after rainy Weather of

 ſome

ſome continuance; particularly, if the Wind blows from any Point of the Compaſs, between the North Eaſt and the North Weſt, at a mean Tide, the Water may be computed to riſe about 10 or 12 Feet perpendicular.

As to the Velocity of the Stream of the River, I made ſeveral Obſervations, from which it may be concluded, that the Velocity of the Surface of the Water, during ſome part of the Tide of Flood, is ſomething greater than at any Time of the Tide of Ebb; the former being near 3 Feet *per* Second, and the other never above 2 $\frac{1}{2}$ Feet *per* Second in this Part of the River.

The Nature of the Ground under the Bed of the River where the Bridge is built, has been examined in ſeveral Places, eſpecially under the Foundation of every Pier, by boring into the ſaid Ground with ſharp and well ſteel'd Drills, urged round, and downwards, by the Force and Weight of ſeveral Men, beſides ſome additional Weights, and it plainly appeared from all thoſe Borings, that there is a very conſiderable Bed of Gravel, quite acroſs the *Thames* where the Bridge is built; that the Surface of this Bed of Gravel, is not at all parallel to the Bottom or Bed of the River; for near the *Weſtminſter* Side, the Gravel is found very near under the ſaid Bed, but the top Part thereof is intermixed in that Place, with ſome thin Layers of Sand, about the Middle the Gravel lies lower, being covered with 3 or 4 Feet of Sand,

Dirt,

Dirt, and Mud; and the Heighth or Thickneſs of this muddy Sand, above the Bed of Gravel, increaſes more and more towards the *Surry* Shore, where the Bed of Gravel lies from 10 to 12 and 14 Feet below the Bottom of the River. The abſolute Depth or Thickneſs of this Bed of Gravel under the Bridge, is certainly very conſiderable, but ſtill unknown. For tho' with a great deal of Time, Patience, and Difficulty, the boring Tool has penetrated in ſome Places 12, and even 17 Feet into that Bed of Gravel, it was found to vary, as to Size and Compactneſs, but in no Place did it ever reach any other Subſtance than Gravel, after the boring Tool had once entered into it, nay, in ſome Places, it was not poſſible to get in the boring Drills above 8 or 10 Feet, and in thoſe Places, when the ballaſt Men dug into it, for laying the Foundations of the Piers, the Gravel was found extremely cloſe, and as it were petrified.

Before I proceed to deſcribe the preſent *Weſtminſter Bridge*, I believe my Readers will not take amiſs, that I premiſe a ſhort Deſcription of the firſt Deſign of the Bridge; in order to which, I muſt inform them, that among the ſeveral Deſigns that were laid before the Commiſſioners, I preſented to the Board, in the Beginning of *May*, 1738, a Deſign of my own for a Stone Bridge, (of which I publiſhed a Plate and Explanation, in *May*, 1739,) conſiſting of 13 large ſemi-circular Arches, ſpringing ſrom about 2 Feet higher than Low Water Mark, 2 ſmaller Arches of 25 Feet each, under each Abutment, and 14 Piers, the greateſt Arch not exceeding 76 Feet in Diameter, all the other large Arches decreaſing by 4 Feet each, on each Side The two largeſt of the 14 Piers, 17 Feet wide, all the other Piers decreaſing, each by one Foot in Breadth, on each Side of thoſe large Piers. The Foundation of each of thoſe Piers, to extend ſeveral Feet all round the Piers, in order to give them a proper Baſis or Footing, and none of them to be laid at a leſs Depth than 5 Feet, below the Surface of the Bed of the River.

The Number and Dimenſions of the Arches and Piers, were approved by the Commiſſioners, and they reſolved, that ſuch Piers ſhould be erected as high, as where the Springing of the Arches ſhould be, that is, 2 Feet or thereabouts above Low-water Mark.

On the 10th of *May*, 1738, the Commiſſioners did me the Honour to appoint me their Engineer, and on the next Board Day after, giving me a general Charge of directing the Works, under the Orders of their Board and Committees only, the Chairman concluded it, by adding, in the Name and by the Direction of the Board, very generous Promiſes, in caſe I ſhould ſucceed in the Methods I propoſed to follow, in building the Piers, which Promiſes have ſince been very fully, and honourably

ably made good to me, without any Sollicitations.

The very firſt Orders that I received, were, *That I ſhould take the utmoſt Care to lay the Foundations of the Piers, and to ſee that the Stone Piers themſelves, ſhould be erected in ſuch a ſolid Manner, of ſuch Dimenſions, and with ſuch Precautions, as might make thoſe Piers capable of ſupporting at any Time hereafter, the Arches, and the Superſtructure of a Stone Bridge, ſuch as my Deſign* (then upon the Table) *repreſented.* The Commiſſioners ordered alſo, at that Time, that over theſe broad Piers, there ſhould be erected little Piers, or Shafts of ſolid Stone, each of theſe to be as long as the Bridge was to be wide, (which was then fixed by the Board at 44 Feet) and 15 Feet in Height, in order to reach ſome Feet above the Surface of the High-water Mark; the Breadth of each of thoſe Shafts, to be 8 Feet, for the firſt 5 Feet in Height; 7 Feet for the next 5 Feet in Height; and 6 Feet for the laſt 5 Feet in Heighth; finiſhing with a large Torus, or *Cordon.* Over thoſe Shafts, the Honourable Board ordered a curiou wooden Superſtructure, of the Invention and Deſign of the late Mr. *James King*, to be erected. As the Model of this Superſtructure, was then ſhown and explained to a great many Perſons of the firſt Rank in the Kingdom, and to many others, and generally approved of, as the moſt curious of its Kind, and the Deſign of it has ſince been

 engraved

Engraved and Printed, I muſt refer the Reader to that Deſign, for further Particulars. What I cannot omit, even in this ſhort Account, is, That by Building theſe Shafts of Stone over the broad Stone Piers, a great deal of Time and Money would be ſaved; and yet, if at any Time hereafter, the Commiſſioners thought proper to finiſh this Bridge with Stone, (which they did in leſs than two Years after) the Arches might be turned, without pulling down any Part of either thoſe Shafts, or of the broad Piers, as it muſt have been done, if the broad Piers, had been continued of the ſame Breadth, quite up to the Height above High-water Mark.

From what has been mentioned, it will appear, that ſuppoſing the Length of the Bridge, or the whole Breadth of the River, 1220 Feet, the greateſt Breadth of all the Piers taken together, did not amount to more than 198 Feet, and the Voids or Openings of all the Arches, would never be leſs than 870 Feet, left free for the Water-way

From theſe Numbers, and the Obſervations of the Velocity of the Stream, mentioned above, I computed the perpendicular Height of the greateſt Fall, that could ever happen under the Arches of ſuch a Bridge, and notwithſtanding the very large Allowances I made, found it could never amount to $3\frac{3}{4}$ Inches, which is ſo inſenſible a Fall, as could prove no Obſtruction at all to the Navigation; and it appeared

peared alſo from Calculations, neither long nor difficult, that ſuch a Bridge could not raiſe the Water in the River at Spring Tides, ſenſibly higher than uſual, or ſenſibly alter the Duration of either Flood or Ebb.

And by the ſame Method, applied to the Circumſtances of *London Bridge*, the perpendicular Height of the Fall, which the Calculation gave, came very near the ſame as I obſerv'd it in the Year 1736, when I found the greateſt Velocity of the Surface of the Stream above the Bridge, at the Time of the greateſt Fall, 3 Feet 2 Inches *per* Second; and the Fall about 4 Feet 9 Inches. And this very Method of Calculation, being adapted properly to the Circumſtances attending *Weſtminſter Bridge*, as it is now finiſhed, it appears, that at that Time of the Ebb, when the greateſt Fall is obſerved, it does not exceed $\frac{1}{4}$ of an Inch, and is certainly leſs than $\frac{1}{2}$ an Inch, there is then no running Water, but only thro' the 13 large Arches, and no Obſtacles to the Stream, but the 12 intermediate Piers, and the Velocity of the Surface of the Stream, juſt above and below the Bridge, is barely 2 Feet and $\frac{1}{4}$ *per* Second; from which Numbers and *Datas*, the Calculation gives, for the perpendicular Height of the greateſt Fall under the Arches of *Weſtminſter Bridge*, about $\frac{3}{10}$ of an Inch, as it really is.

I cannot conclude this Article, without obſerving, that it is no Wonder, that the pretended ed Calculations of the Fall under *Weſtminſter Bridge*, publiſhed in ſeveral Pamphlets, ſhould be all falſe; ſince the Authors never thought at all, or inquired, about the greateſt Velocity of the ebbing Waters, or ever had taken that Velocity (on which the Fall chiefly depends) into any of their Calculations; nay, ſome of them were ſo intirely ignorant of the very firſt Principles on which thoſe Computations are founded, that they attempted very gravely, to compute and compare the Fall under *Weſtminſter Bridge*, with that under *London Bridge*, by the Rule of Three; by which alone, they might as well attempt to find the Moon's Place, or the preſent Value of the Reverſion of an Eſtate, after two or three Lives. The Tides, and all other Circumſtances attending the *River Thames*, and the Nature of the Ground under its Bed, being duly conſidered, the Method of laying the Foundations of the Piers, which I thought moſt likely to ſucceed, was that which I had the Honour of explaining twice upon working Models, before the Commiſſioners, a ſhort Account of which is as follows.

That the Foundation of every Pier ſhould be laid on a ſtrong grating of Timber, planked underneath. --- That this Grating of Timber ſhould be made the Bottom of a Veſſel, ſuch as is called *Caiſſon*, by the *French*; (which Term I ſhall make Uſe of) That the Sides of this *Caiſſon*, ſhould be ſo contrived, as to be taken away, after the Pier ſhould be finiſhed: That the

the Bed of the River ſhould be dug to a ſufficient Depth, and made level, in order to lay thereon the Bottom of the *Caiſſon*: That where-ever the Ground, under the Excavation, or Pit, ſhould prove good, there would be no Neceſſity for piling it; but that in Caſe the Ground under the Foundation Pit, ſhould not prove of a ſufficient Conſiſtence, it ſhould be piled all over as cloſe as neceſſary, the Heads of thoſe Piles ſawn level, cloſe to the Bottom of the Pit, and on the Tops of thoſe Piles, the Grating, and Foundation of the Pier, ſhould be laid as is uſual in ſuch Caſes; and there is nothing uncommon, or very difficult, in performing theſe Operations, tho' always attended with a conſiderable Expence of Time and Money. For Piles are daily, and have been driven and ſawn under Water, in many Places, almoſt as faſt, and full as well as above Water; and I believe no where better or faſter, than by the Machines which I recommended, and were employed for thoſe Purpoſes.

This may ſerve as a general Idea of the Method of laying the Foundations of the Piers; but thoſe Readers who require a further Detail, will find it at the End.

I ſhall now proceed to give my Reader a ſhort Deſcription of *Weſtminſter Bridge*, as it was ordered into Execution, by the Commiſſioners, on the 12th of *March*, 1739-40, previous to which, I think it proper to mention, that the Methods I had propoſed, to lay the

Foun-

Foundations of the Piers, had already ſucceeded ſo well, in the Execution, that in leſs than 11 Months, *viz.* from *January* 29, 1738-9, to the 3d Day of *December* following, we had compleatly built and finiſhed the 2 largeſt Piers, to above High-water Mark, and another Pier to Low-water Mark. But the great Froſt which then followed, put an entire Stop to the Works of the Bridge, and tore to Pieces, or carried away all the Piles and other wooden Work, then ſtanding in the River. During that Interruption, ſome Commiſſioners obſerved at the Boards, that the Goodneſs of the Method made uſe of in Building the Piers, was then ſufficiently tried. That the Publick in general, was highly diſguſted, at the Thoughts of having a Wooden Bridge, in the Metropolis of the *Britiſh* Empire, no ways ſuitable to the fine Tide River, over which it was to be built: That if ſuch a wooden Bridge was ever finiſhed, it would be in great Danger of being carried away, or greatly damaged, by any future Heaps of Ice, ſuch as were then on the frozen *Thames*; and that at beſt, ſuch a wooden Bridge would, in a few Years, amount to the Expence of a laſting Stone Bridge, on account of the frequent and expenſive Repairs, it muſt neceſſarily be in want of; which being well conſidered, the two Maſter Carpenters, who had contracted for building the wooden Superſtructure for 28,000 *l.* were deſired in *February*, 1739-40 to declare, whether they would undertake

dertake to keep that Superſtructure in Repair for a certain Number of Years, and for what Sum. But (after having taken Time to conſider) they gave as their Anſwer, that they would build the ſaid wooden Superſtructure, according to their Contract, but did not care to undertake, on any Terms, to keep it in Repair afterwards. This honeſt Anſwer, joined to the Reaſons mentioned above, determined the Commiſſioners to drop the Deſign of a wooden Bridge, and they reſolved upon an intire Stone Bridge, on the 23d of *February*, 1739-40, for which I was ordered to prepare a Deſign, ſuited to the Number and Sizes of the Piers already reſolved upon; but the Stone Bridge was not ordered into Execution, till the Carpenters had received a conſiderable Sum of Money, as a full Satisfaction, to indemnify them, for what they had laid out, and a very conſiderable Profit upon the whole Sum of their Contract, which they then gave up in order to be cancell'd, as well as the Bonds entered into by their Securities.

This new Deſign of a Stone Bridge, I preſented to the Board of Commiſſioners on the 12th of *March*, 1739-40. It does not differ in any eſſential Part, from that which I had offered to them two Years before. The Number, Figure, Size, and Dimenſions of its ſeveral Piers, Abutments, and Arches, being ſtill the ſame, as well as the Choice, and Diſpoſition of the ſeveral Materials uſed in its Inſide

and

and Outſide, all which had been generally well approved of, but as to the Decorations of the Points of the Piers, the Turrets over them, the Corniſhes, Foot Ways, Receſſes, Plinths, and Balluſtrades, the new Deſign had been much better conſidered, and even ſince, all theſe have been greatly improved, both as to Conveniency and Beauty, by the many uſeful Hints, and Advices, which from Time to Time I received, from the Right Honourable the late Earl of *Pembroke*, &c. to whoſe noble and publick Spirit, conſtant Care, and ſingular Diſintereſtedneſs, the Publick chiefly owes, not only the having a Bridge at *Weſtminſter*, but alſo the having it (perhaps) the beſt built, and certainly one of the moſt magnificent Bridges in the whole World.

The whole Length of the Bridge, that is to ſay, the whole Breadth of the *River Thames*, from the *Woolſtaple Dock*, to the oppoſite Wharf, being 1220 Feet, is diſtributed into 13 large Arches, 2 ſmaller Arches, 14 Piers, and 2 Abutments, of the following Dimenſions.

FEET.

	FEET.	
Weftminfter	76	Abutment
Abutment Arch	25	
	12	Abutment Pier
An Arch	52	
	12	a Pier
An Arch	56	
	13	a Pier
An Arch	60	
	14	a Pier
An Arch	64	
	15	a Pier
An Arch	68	
	16	a Pier
An Arch	72	
	17	the Eaftern Middle Pier
The Middle Arch	76	
	17	the Weftern Middle Pier
An Arch	72	
	16	a Pier
An Arch	68	
	15	a Pier
An Arch	64	
	14	a Pier
An Arch	60	
	13	a Pier
An Arch	56	
	12	a Pier
An Arch	52	
	12	Abutment Pier
Abutment Arch	25	
Surry	76	Abutment
Whole Breadth	1220	of the River Thames

Length of the 2 Abutments,	152	} Solids — 350 Feet
Section of the 14 Piers, —	198	}
Span of the 15 Arches, —	870	Voids — 870
Section of the River Thames,		—1220 Feet.

The whole Breadth of the Bridge, as it was fixed by the Honourable Board, is 44 Feet, from out to out; the Top of the Bridge is divided into three Walks, the Middle one 28 Feet in Breadth, for the Horses, Cattle, and Carriages, which is more than sufficient for three Carriages and 2 Horses besides, a-breast.

The side Walks for the foot Passengers, are raised about a Foot above the Carriage way, the Breadth of each, being near 7 Feet in the Clear, as may be seen at the Bottom of the annexed Plate, which represents the Plan of one half of the Top of the Bridge, and the Plan of one Half of the Foundations of the Stone Piers; where may also be observed, the Extent of the Beds, or Gratings of Timber, (upon which, each of the Piers is built) all round the said Piers.

The little Squares shew the Piles, which were driven to support the wooden Centers, on which the Stone Arches were turn'd, which Piles (after the Centers were eased, struck, and removed) have been all sawn off several Feet lower than the Low-water Mark, so that the Vessels of the deepest Draught of Water, that are navigated above *London Bridge*, can never touch any of them, or receive Damage from them.

Each of the Piers is terminated by a right Angle at each End, as sufficiently sharp to divide the Waters in so gentle a River as the *Thames*, and the most proper Angle to make good Work; and their Outſides on both Fronts, are decorated in the Shape of Pedeſtals. And in order to make thoſe Piers the ſtronger, the two loweſt Courſes in each are two Feet high each, with an Offset, or Retreat of one Foot each, all round, like two Plinths, or Steps, one over the other. The Foundation of each of the Piers, has been laid at leaſt, 5 Feet below the Surface of the Bed of the River, and lower, where neceſſary, as may be ſeen in the annexed Plate.

The upper and lower Plans of one of the two Abutments, (as they are now built) are alſo truly repreſented in the annexed Plate; each of the Land Breaſts ſpreads about 25 Feet on each Side of the Bridge, in order to ſtrengthen the whole Fabrick, and to afford Room, always wanted at the Foot of all conſiderable Bridges.

Theſe Abutments have ſeveral Advantages; among others, the Stairs and Cauſeways are properly placed, for the Conveniency of Water Paſſengers; and the loading and landing of Goods, will be at all Times out of the Indraught of the Arches, beſides convenient Room for the Boats, and for the Watermen to ply for Fares, without embarraſſing the Streets, leading to and from the Bridge. Laſtly, theſe Abutments may, in time, lead the Way to the making of moſt uſeful and beautiful Keys along the River, between High and Low-water Mark, than which nothing can more contribute to the Trade and Ornament of the City

 and

and Liberty of *Westminster*, and to the Preservation and Improvement of the Navigation of the River, which would thereby, have always a sufficient Stream, to clear its Bed from Sand, Mud, and Shoals, and would always retain Water enough, for working and navigating of Boats, and other Crafts or Vessels, and for the Loading and Unloading of them, at all Times, with Ease. The great Advantages and Conveniencies attending such Keys, may be observed in those which the City of *London* has most prudently advanced into the River: For there they can work, and carry on Business a much longer Time each Tide than at *Westminster*, and the opposite Shore, which are left useless, bare, and covered with unhealthy Ouze and Filth, the greatest Part of every Tide.

The upper Part of the annexed Plate, is taken up with the geometrical Elevation of one of the Fronts of *Westminster Bridge*, and its Abutments, under which may be observed, a true Section of the River, at Low-water; the Surface of High-water, at the highest Spring-Tides, represented in the Arches, by a prick'd Line; the Depth of the Bed of Land and Mud, over the Bed of Gravel, in which the Piers are all built, represented in a Layer of small Dots; as also the Depth of the Foundations of every Pier, Abutments, and Water-Stairs, the Position of the Rows of Piles; that were drove

for

for the Centers, and the Situation and Extent of the Beds of Timber under the Piers.

I beg Leave to add a few Words, in order to ſhew it was not altogether Fancy, or any Affectation of a particular Taſte and Stile, that made me chuſe this Sort of Elevation for the Fronts of *Weſtminſter Bridge*.

The Proportion of the Piers to the Arches, in this Deſign, (which in Bridge-building, ought always to be the firſt Conſideration) is ſuch as gives a very great Strength to the whole Fabrick; for the moſt approved Authors give to the Piers only $\frac{1}{6}$ or $\frac{1}{5}$ of the Span, or Opening of the Arches, as ſufficient; whereas, the Piers built are ſuch, that the Arches are hardly more any where, than 4 Times and $\frac{1}{2}$ wider than the Piers that ſupport them; that is, the Width of the Arches are to the Width of the Piers that ſupport them, nearly as 9 to 2, and yet more free Water-way is left for the Stream of Flood and Ebb to paſs under it at all Times, than the whole Breadth of the River at the *Horſe-ferry*, and very near as much as the whole Breadth of the River at *London Bridge*.

Theſe Piers are not built in the common and uſual Way that is practiſed in the building of them, *viz.* to make an outward Shell of hard Stones, regularly cut, and ſet in Courſes, and all the Inſide fill'd with Brick-work, or common Rubble; but they are all built entirely ſolid, the ſame in their Inſide as their Outſide, with large Blocks of *Portland* Stone, laid

in

in regular Courſes, all the Joints fill'd with a Cement, made of Lime and *Dutch Tarris*, which ſets and hardens in Water, and the Stones of every Courſe cramped together with Iron Cramps, let into the Stones, and runn'd in with melted Lead; and thoſe Cramps are ſo placed, that not one of them can ever be ſeen, or be affected by the Water.

The Arches are all ſemi-circular, not only as ſtronger than any Elliptical, or than any Segment of a Circle of the ſame Span; but alſo becauſe their Centering and Execution, are leſs liable to Difficulties, and that moſt People look upon the Semi-circular Arches as the moſt graceful.

All the Arches are built, ſo as to ſpring from two Feet, or thereabouts, above the Level of Low-water Mark, and from no higher, for many Reaſons: 1*ſt*, A great deal of Stone Work, Time, and Expence, is thereby ſaved. 2*dly* The Arches are much ſtronger, and their *Thruſt*, or *lateral Preſſure* much leſs, by being placed upon ſuch low Piers. 3*dly*, The Aſcent of the Bridge is much eaſier, and leſs fatiguing for Men and Cattle. 4*thly*, The Abutments of the Bridge do thereby extend but a little Way; whereas, in all the other Deſigns I ever ſaw for this Bridge, the Arches are made to ſpring from High-water, or thereabouts, and conſequently would be much weaker, their *Thruſt*, or *lateral Preſſure* much greater, and the Expence of building the Piers much increaſed, without the

the leaſt Neceſſity. Beſides, there would be no eaſy getting on or off of ſuch high Bridges, unleſs the Abutments on each Side of the River reached ſeveral hundred Feet upon the dry Land.

The *Coins*, or *Vouſſoirs*, or Arch Stones, in each Front, have chamfer'd Joints; which are continued quite thro', under the *Suffete* of all the Arches, becauſe Experience ſhews this ruſtick Decoration, of the *Archivolt* of the Arches, to have a very good Effect, in large Works eſpecially; and that the chamfering the Joints, hinders the fluſhing, or breaking of the Edges of the Stones.

In order to give the utmoſt Strength to the Arches of the Bridge, I deſign'd their Conſtruction very differently from the common Way of building ſuch Arches; for in order to deſtroy, or counterballance the *Thruſt*, or *lateral Preſſure*, with which all Arches (even the ſemicircular ones) do endeavour to ſeparate, or overſet their Piers, every Arch of *Weſtminſter Bridge* (except the two ſmall ones at the Abutments) is double, the firſt Arch is ſemi-circular, built with great Blocks of *Portland* Stone, from 3 to 5 Feet in Height, or Depth; over which there is another Arch built with Purbeck Stones, bonded in with the under ſemicircular Arch. This upper Arch, is of a particular Figure, or Curve, four or five Times thicker in the Reins, or towards the Bottom, than at the Key or Top. Both theſe Arches taken together, do form a Kind of Arch, which can be demonſtrated to be *in equilibrio*, in all its Parts: By means of theſe ſecondary Arches, and the proper Diſpoſition of the ſuper-incumbent Materials, every Arch of *Weſtminſter Bridge* is able to ſtand by itſelf, independent from the Abutments, or any other Arch. I aſſerted above 12 Years ago, that Arches thus conſtructed, muſt have that Property, as a neceſſary Conſequence, from a Mathematical Propoſition, as clearly demonſtrated as any one Propoſition in *Euclid* or *Apollonius* and the Truth of my Aſſertion has ſince been put out of all Doubt; for when by the ſettling of the Weſtern 15 Foot Pier, in 1747, it became neceſſary to take down the two adjoining Arches, and to rebuild them, all the other Arches, even the next to them on each Side, ſtood firm and well, (tho' unſupported on one Side) nor were they at all affected, by two ſevere Shocks of Earthquakes, that were felt in *London*, in *Februry* and *March* 1749, to the great Amazement of many, and the no leſs Confuſion and Diſappointment of not a few malicious or ignorant People, who had confidently aſſerted, and propagated the Notion, that upon unkeying any one of the Arches the whole Bridge would fall.

Between every two Arches, I have managed proper Drains, to carry off the Rain and other Waters, which might, in Time, accumulate in thoſe Places, to the great Detriment of the Arches;

Arches; ſome Bridges having been ruin'd for want of this Precaution, which ſhould be obſerved in all conſiderable Stone or Brick Bridges, and yet is to be found in very few.

As to the Fronts of the Spandrells of the Arches, they are filled with good and regular Purbeck Stones, with proper Bond, and the Joints of the Work preſerve a Tendency to the Center, as is expreſſed in the annexed Deſign. This Manner of filling the Spandrels of the Arches, is much preferable to the common Way, which is, to fill what is above the Arch Stones, in the Fronts, with horizontal Courſes of Stone or Brick, and to fill all the Inſide with Rubble, laid at random.

It is ſurprizing, that this Manner of Arching has not been put in Practice ſo often as it might. However, I ſhall mention a few Examples, where the ſame good Precaution has been obſerved: The great Arch at *Venice*, call'd the *Rialto*, near 100 Feet ſpan; the great Arch at *Vicenza*, of above 100 Feet ſpan; a Groin Arch at *Blenheim*, built in this Manner, with Rubble Stones only, which ſtands firm and well, though it has only 3 Feet $\frac{1}{2}$ Riſe, upon 44 Feet ſpan; and all the Arches of the *Pont-Royal*, at *Paris*, ſo very much cried up by the *French*, and ſo juſtly to be praiſed, on this Account, at leaſt.

Over each Point, or ſaliant Angle of each of the Piers, there is a Semi-octogonal ruſticated Turret, built with Stone, for the follow-

 ing

ing Reaſons: In order, in the *firſt* Place, to have the Points, and the Middle of the Piers, as equally loaded as poſſible, which will very much contribute to the Security of the whole. *2dly*, To ſtrengthen the Arches, by oppoſing ſo much more Weight or Reſiſtance againſt their *Thruſt*, or *latural Preſſure*; for it can be demonſtrated, that the lighter an Arch is, in Proportion to its Piers, or (what comes to the ſame) the heavier the Piers are, in Proportion to the Arch, the firmer the Arch will be; and the contrary vulgar Opinion, *viz. That the more an Arch is loaded, the ſtronger it will be*, is a groſs Error, as may eaſily be ſhewn. *3dly*, Theſe ruſticated Turrets, beſides the real Advantages already mentioned, do very much add to the Decoration of the Fronts of the Bridge, by dividing, or breaking ſo long a Line as the whole Length of this Bridge, into as many Parts as there are Arches. *4thly*, Becauſe theſe Turrets being carried quite up, and the ruſtick Corniſh, and the Parapet Walls and Balluſtrade, made to follow their Out-lines, they afford uſeful and commodious Receſſes for the Foot Paſſengers; where they may retreat, if any Buſineſs, or Accident requires their Stopping, without embarraſſing the Foot-ways, as it happens but too often in the Streets.

I know not how to finiſh this ſhort Deſcription of *Weſtminſter Bridge*, as it is now finiſh'd, better than by obſerving,

That

That the Want of another Bridge, besides that at *London*, in the Capital of the *British* Dominions, had been felt long ago:

That Attempts to obtain such a Bridge have been made in the several Reigns of Queen *Elizabeth*, *James* I. *Charles* I. *Charles* II. and *George* I. and constantly defeated, by Means too well known to require my mentioning them:

That publick Good, and publick Spirit, having (in this Case) got the better of private Interest, *Westminster Bridge* was petitioned for, obtain'd, undertaken, begun, and finished, under the Reign of his present Majesty: That this Bridge was built without turning of the whole, or any Part of the River; without stopping, or even hindering the free Navigation one single Moment, and without having any sensible Fall under its Arches:

That this Bridge has certainly nothing of the Kind in *Europe*, and perhaps in the whole World, that can be brought in Competition with, and much less exceed it; especially, if its Situation and Construction are considered, *viz.* in such a Metropolis as *London*, extending over so wide a Tide River, as the *Thames* in this Place, and built upon Piers of solid Stone, laid several Feet below the Surface of its Bed:

That this Bridge will be in no Danger of being destroyed, by the Action of the Tide and the Ice, in any hard Winter, which soon or late, is the Fate of all Bridges, built upon wooden

wooden Piles, in Rivers that carry Ice in Winter, and where the Tide riſes to any conſiderable Height:

That this Bridge, (its Paving and Gravelling excepted) will want no conſiderable Repairs for a long Courſe of Years, and be in no Danger of being damaged by violent Storms of Wind, ſet on Fire by Lightning, or other Accidents; or maliciouſly burnt down by evil minded People:

That this Bridge is not only a beautiful, but a moſt laſting, uſeful, and neceſſary Communication between the neighbouring Counties; a conſiderable Means towards the Increaſe of Trade, Manufactures, and uſeful Arts; a very great Ornament to the Capital of the *Britiſh* Empire; an Honour to the Commiſſioners in particular, and to the whole Legiſlature in general, who have not only given Sanction to the building of this Bridge, but provided in ſo ample a Manner the neceſſary Funds; and laſtly, an Event not unworthy of being recorded in the Annals of his preſent Majeſty's moſt glorious and happy Reign

An

An Account of the Methods made Uſe, of in laying the Foundations of the Piers of Weſtminſter Bridge; *with an Anſwer to the chief Objections that have been made thereto.*

THOSE of my Readers, who want only a general Idea of the Manner in which the Foundations of the Piers were laid, will find ſuch an Account in Pages 10 and 11; but for thoſe who chuſe to enter into a greater Detail, I have reprinted the following Account of the moſt material Tranſactions at the Works of the Bridge, from the Time that the two largeſt Piers were ordered into Execution, (*viz.* from the End of *June* 1738) to the Time the firſt Pier was finiſhed.

About the Beginning of *July* 1738, I went in Company with the Maſons, to the Iſland of *Portland*, where I took a View of the Quarries, and made ſuch Inquiries as I thought might be of Uſe hereafter. The Maſons did immediately ſet a ſufficient Number of Hands (moſt of them ſent down there from *London*) upon cutting out of the Quarries, and working into proper Scantlings all the Stone neceſſary for the two largeſt Piers.

The

The Month of *July* following was ſpent in getting ready, and preparing Timbers, making neceſſary Tools, and in conſtructing an Engine to drive Piles; of which I ſhall ſay ſomething more in its proper Place.

In the next Month of *Auguſt*, a ſmall Spot of Ground was procured for the Carpenters to work in, on the *Surry* Shore; and tho' ſome Months after we got Poſſeſſion of a ſmall Wharf, and Crane adjoining, we ſhould have been very much ſtreighten'd for want of Room, and proper Conveniencies, if it had not been for the Kindneſs and Civility of ſeveral Gentlemen of that Side of the River, who freely offered us the Uſe of whatever they had, that could be of Service to us,

About this Time the Carpenters began to make and erect, under the *Surry* Shore, 12 Frames of Timber, ſupported in a vertical Situation, parallel one to another, and kept in their Places by ſhort Stakes or Piles, driven into the Ground. Theſe Frames reached about 2 Feet above the common High-water Mark, and were braced together, ſo as to be kept upright and ſteady, till the *Caiſſon* ſhould be built and finiſhed, on the Top of them. The Ground-cills, or bottom Pieces of theſe Frames were made cylindrical, that by taking the Braces away, they might ſerve as ſo many Axes, or Center-pieces, round which the Frame could move, or revolve altogether, for the lowering or launching of the finiſhed *Caiſſon*, at any High-Water, without

without any Danger of racking or ſtraining. The Grating for the firſt Pier, was alſo begun this Month, and the Engine for driving Piles being almoſt finiſhed, I directed that it ſhould be framed upon the Sides of two Barges, adapted to that Purpoſe, in which, a conſiderable Quantity of large Ballaſt was ſtowed, to make this Engine, and its floating Stage, as ſteady as poſſible.

In the Beginning of *September*, the Engine for driving Piles was finiſhed and tried; and whilſt the Carpenters were at Work on the Grating and Planking for the *Caiſſon* of the firſt Pier, we began to drive Firr Piles, of about 13 or 14 Inches ſquare, and about 34 Feet in Length; the Points of thoſe Piles were ſhod with Iron, in a particucular Manner, to penetrate the eaſier into the hard Gravel, and the Tops of them were fortified with a large and thick Iron Hoop, which was taken off, to ſerve to the ſame Uſe again, after we had driven the Piles as low as we could, without Danger of breaking or ſplitting them, which was about 13 or 14 Feet below the Surface of the Bed of the River. Theſe Piles were placed about 7 Feet aſunder, in Lines parallel to the two ſhort Sides of each End of the intended Pier, and at about the Diſtance of 30 Feet from them; the Uſe of theſe *Fenders*, or Guard Piles, was to ſecure the Works from the Approach of Barges, and other large Veſſels;

and

and as to other Boats, they were hindered from paſſing between the Piles, by long Pieces of Timber, called *Booms*, which floated up and down with the Tide, alongſide the Piles, to which they were faſtened by wooden Rings, and proper Iron-work. By means of theſe Booms, we could incloſe the Works, and all our own Boats and Veſſels, from being damaged or rummaged, either by Day or Night, without taking up but a very ſmall Part of the River.

On the 13th of *September* 1738, we drove the firſt Pile, being the Southermoſt of thoſe which guarded the upper Point of the Weſtermoſt large Pier; and we continued driving Piles when the Weather was not too bad, or nothing more preſſing obliged us to employ the Workmen in other Things; and on the 26th of *October*, the Piles neceſſary for building the two middle Piers, were all driven, and found to ſtand ſo firmly, that all the Plates, Whale-pieces, Ties, and Braces, that had been contrived to keep them ſteady, (in caſe they ſhould have wanted it) were omitted, as entirely ſuperfluous, which ſaved a great deal of Time and Expence.

The Number of Piles which were neceſſary for Building the firſt large Pier, according to this Method, was only 34 long ones, and 22 ſhort ones; and for the other large Pier, 26 long ones only: Whence thoſe that are acquainted with theſe Sorts of Works, will perceive

ceive how much Time and Expence was ſaved, by proceeding in this Manner. Whereas, if the building of this one Pier had been attempted with a Coffer-dam, or *Battardeau*; (for the Conſtruction of which, inſtead of 56, there would have been a Neceſſity of driving about a thouſand Piles, beſides the Expence of Clay, Boards, and a great Number of croſs Beams and Braces to ſupport it, &c.) the ſaid Coffer-dam, or *Battardeau*, could never have been drained of the Water perpetually ouzing up thro' ſuch a gravelly Soil, as is the Bed of the River in this Place. Nay; if the Soil of the River had been a Clay, and would have allowed of *Battardeaux*, the Building of Piers in ſuch a Manner would have been much leſs expeditious, and much more coſtly. I ſhall ſay ſomething more on this Subject in its proper Place.

The Method made uſe of to drive the Piles, was contrived by the late Mr. *James Vauloüé*, a very ingenious Watch maker of my Acquaintance, who has publiſhed a Print of the Engine, with an Explanation; for which Reaſon, it will be ſufficient for me to mention, that having viewed the Model of that Contrivance, and calculated the Effect of ſuch an Engine, I found that ſuppoſing the Ram or Weight to be 1700lb. and the Height of the Strokes at a Mean, 20 Feet perpendicular, the Engine would give about 48 Strokes *per* Hour, by the Help of two Horſes, and about 70 Strokes *per* Hour, by the Help of

three

three Horſes: This Effect being much ſuperior to that of any of the Engines, commonly uſed for that Purpoſe, was the chief Reaſon that determined me to drive our Piles in that Manner; for though the Expence of our Engine was conſiderably more than that of any other Sort, and though it took a great deal of Time in making, yet I knew it would make us Amends by the Expedition, and other Conveniencies attending this Manner of driving Piles upon a floating Stage, as we afterwards found it did.

I alſo directed ſome Contrivances, deſcribed in the Print of our Engine, to be omitted, as unneceſſary in the Engine itſelf, in order to have it intirely under Command at every Stroke (as all ſuch Engines ſhould be) and to avoid Danger and Miſchief as much as poſſible.

After our Engine had worked long enough to have the *Gudgeons*, or *Pivots*, and all the rubbing Parts made ſmooth, and the Stiffneſs of the Ropes in a great Meaſure deſtroyed; Part of the Friction which I had conſidered, and allowed for in calculating the Effect, was removed, and then the Engine performed ſomething better than according to the Calculation: So that by the Force of three Horſes going at a common Pace, when the Height to which the Ram was raiſed did not exceed 8 or 10 Feet, the Engine gave about five Strokes in two Minutes.

About

About the Beginning of *October*, the Grating belonging to the first Pier was finished, and the Sides of the *Caisson* 16 Feet in Height, began to be raised upon it. These Sides were made of Fir Timbers, laid horizontally and close one over another, pinned with Oaken Trunnels, and framed together at all Corners of the *Caisson* except the two Points, or saliant Angles, where they were secured by proper Iron Work; which being unskrewed, would permit the Sides of the *Caisson* (in case it should be found necessary) to part asunder into two Halves. These Sides were planked across the Timbers, inside and outside, with three Inch Planks in a vertical Position. The Thickness of those Sides was 18 Inches at Bottom, and 15 Inches at the Top: And, in order to be further strengthen'd, every Angle but the two Points had three Oaken *Knee Timbers*, properly bolted and secured. These Sides when finished, were fastened to the Bottom or Grating, by 28 Pieces of Timber on the Outside, and 18 withinside, called Straps, about 8 Inches broad, and about 3 Inches thick, reaching and lapping over the Top of the Sides; the lower Part of these Straps had one Side cut *Dovetail Fashion*, in order to fit the *Mortoises* that were cut to receive them in the outermost Kerb of the Grating, and kept in their respective Places by Iron Wedges. The Use of these Straps and Wedges is as follows.

When the Pier is built up so much above Low-water, as no longer to want the *Caisson*

 for

for the Masons to work in, the Wedges being drawn up, give Liberty to clear the Straps from the Mortoises; the Consequence of which is, that the Sides must rise by their own Levitation or Buoyancy, leaving the Grating under the Foundation of the Pier, which Grating extends several Feet all round, by way of Footing or Basis, as is, or should be always observed in such Buildings.

The Sides of the *Caisson* were hindered from being pressed in by the ambient Water, towards the Stone Work that was to be placed in it, by Means of a Ground Timber or Ribbon 14 Inches wide, and 7 Inches thick, pinn'd upon the upper Row of Timbers of the Grating, which exactly fill'd the Space contain'd between the Sides of the *Caisson*, and the first Course or lower Plinth of the Stone Pier; and the Top of the Sides was secured by a sufficient Number of Beams laid a-cross, which also served to support a Floor, on which the Masons and Labourers stood to hoist the Stones out of the Lighters, and to lower them into the *Caisson*: Several other Frames of Timber had been contrived to keep the Sides from being crushed inwards; but those Sides alone proved strong enough, and all the rest were omitted, excepting some few Struts or Props, placed occasionally between the Sides and the Stone Work, which saved both Time and Expence.

Towards the End of this Month of *October*, the Ballast-men began to make the Excavation, or Foundation-Pit for the Westermost of the two large Piers. They had Orders from me to dig it near 6 Feet in Depth, and the Dimensions of the Pit to be in the same Shape as the *Caisson*, and 5 Feet wider all round, with a Slope sufficiently gentle to hinder the Ground from falling in again; and, in a great Measure, to prevent the loose washing and silting of the River from coming into the Pit, short groved Piles reaching not above four Feet higher than Low-water Mark, were driven before the two Ends, and Part of the Sides of the intended Pier, in Lines parallel to the *Fenders or Guard Piles*, and at about 15 Feet Distance from the Sides of the *Caisson*: Then two Rows of Boards were let into the Groves down to the Bed of the River, and kept from rising by Ledges of Wood nailed in the Groves of the short Piles.

In the Beginning of *January* following, which was some Time after the Digging the Pit, and leveling its Bottom was finished; I examined the whole Surface very carefully in the Presence of several Persons, and it appeared that the Cavity had been dug according to the Dimensions I had given, and that no Part of the Bottom was more than one Inch or thereabouts from a true Level, which is a much truer Level than what is necessary, or what is observed in laying the Foundation of any Building. The Digging the Foundation regularly, and to a sufficient Depth, wat directed by the Help of a great Number *of Gauges*, each of which consisted of a Stone about 15 Inches Square, and 3 Inches thick, in the Middle of which was fixed a wooden Pole or Stem, 18 Feet in Height, divided into Feet and Inches from the Bottom of the Stone, the Pole painted red, and the Numbers and Divisions white, the easier to distinguish when the level Surface of the Water intercepted equal Portions of those *Gauges*. -- The Examination of the Bottoms Level, was also performed by the Help of one of those *Gauges*, which was applied to every Part of the Bottom successively, in so regular a Manner, as to be certain not to leave one single square Foot in the whole Surface of the Foundation untried.

About this Time the *Caisson* was finished, with a Sluice towards its Bottom; and all the necessary Iron-work, to moor or fasten it by, all the Seams or Joints caulk'd, and the Bottom and Outside pay'd or pitched over.

On the 15th of *January* 1738-9, the *Caisson* was launched very successfully without racking, or touching the Sides of the Wharfs, along which it moved, which to prevent, a Lighter had been strongly moored in the River, at about 200 Feet Distance from the Shore, over against the Place where the *Caisson* would be as soon as it reached the Water; from the Head and Stern of this Lighter, two Cables were stretched to the two Ends of the *Caisson*, whereby its Motion in the Water, after it was launched, proved exactly as it was intended, without any Person on board to direct it, or tow it, till the said *Caisson*

Caiſſon was near the Piles that ſurrounded the Place where it was to be ſunk.

On the 29th of *January*, the Maſons began hoiſting ſome Stones into the *Caiſſon*; and in the Afternoon, *the Right Honourable the late Earl of Pembroke*, &c. condeſcended to ſet the firſt Stone, which was the Middle Stone of the Foundation of the firſt Pier. Soon after which, his Lordſhip retired, leaving ample Marks of his wonted Generoſity among the Workmen.

In the Beginning of *February*, whilſt the Carpenters were employ'd in preparing Timbers, and framing a new Grating or Bottom, for the Foundation of another Pier, the Maſons continued in hoiſting and ſetting the Stones of the firſt Courſe which being finiſhed, by lifting the Gate of the Sluice, near the Time of Low-water, we ſunk the *Caiſſon* with the Stones in it to try how it ſet and grounded.

By this firſt Trial, we found ſome looſe Ground had tumbled into the Pit, (which was chiefly occaſioned by a Barge, that had been maliciouſly ſunk, ſo as to hang in Part over the Bank or Slope of the Pit) upon which the Sluice was ſhut again, and in leſs than two Hours pumping, we made the *Caiſſon* float as before, and drain'd all the Water that had been let into it.

The two next Days, the looſe Ground that had tumbled in, was taken out; and the whole Surface of the Bottom of the Pit was examined again, and found as level as when firſt finiſhed.

Soon after the Maſons cramped the Stone of the firſt Courſe, and having ſet and cramped the ſecond Courſe, by lifting the Gate of the Sluice, we ſunk the *Caiſſon* a ſecond Time, and found it to bed itſelf, or ſet perfectly Level up-the hard Gravel. Being now entirely ſatisfied with the Level of this Foundation, we pumped the Water out of the *Caiſſon* again, and made it float as before.

About this Time the Ballaſt-men began the Excavation for the Foundation of the ſecond Pier, which was encloſed all round to hinder the waſhing of the River from coming in, the two Points by three Rows of Boards ſet edge-ways, one over the other, (the lowermoſt down to the Ground) faſtened in a proper Manner to the guard Piles themſelves, which ſaved the Expence and Time of driving ſhort Piles; and the long-Sides of that Incloſure were ſecured by ſome long Troughs fill'd with Gravel, and let down upon the Bed of the River, reaching from one ſhoulder Pile to the other, on each Side.

In digging this Foundation, there was found a Copper Medal, about the Size of a Half-penny, tolerably well preſerved; the Head of the Emperor *Domitian* on one Side, as appears by the Inſcription round it; the Reverſe, a Woman ſtanding, with a Pair of Scales in her Right Hand, and ſupporting a *Cornucopia* with her

her Left, and these Letters round, MONETA AVGVSTI, with the initial Letters of SENATVS CONSVLTO; the S under the Scales, and the C on the other Side of the Figure. This Medal looks very much like a cast one, tho' it is probable it is a genuine one, because lying so long under Water might give it the Appearance it has; besides, Copper Medals of this Emperor, and of the same Sort of Reverse, are so common, that it would not be worth while to counterfeit them: As to its being found there, it is easily accounted for, if it be true, that there was a Ferry about this Place, in the Time of the *Romans*, and there are many things which confirm this Opinion.

The Masons proceed d in setting and cramping the third Course of Stones; which being entirely finished on the 16th of *February*, I made the proper Observations, in order to place the Pier at Right Angles, with the middle Lines of the intended Bridge, and on the 17th the *Caisson* was suffered to sink for the last Time; for it was found to bed or ground so nearly true, as to Situation and Level, as not to need to be altered in the least.

The Stone Work of the Pier, was at that Time brought up within about two Feet of the common Low water Mark, or in other Words, the Building of the Pier was then brought *within the reach of the meanest Capacity:* For the Remainder being only common Tide-work, has nothing worth relating, and a particular

 Account

Account of it would be foreign to my preſent Purpoſe, which was to give (according to Order) a ſhort Account of the Method uſed in laying the Foundation only. However, as this Account may fall into the Hands of People, utterly Strangers to what is meant by Tide-work, I will, in few Words, explain the Manner of our Proceeding every Tide, at leaſt, till the Pier being built up ſome Feet above Low-water Mark, the Maſons made uſe of the *Caiſſon* no longer.

About two Hours before Low-water, in order to get Time, the Sluice of the *Caiſſon*, kept open till then, (leſt the Water flowing to the Height of many more Feet on the Outſide than in the Inſide, ſhould float the *Caiſſon*, and all the Stone Work out of its true Place) was ſhut down; and by the Help of only four Pumps of 8 Inches Square, three Men to each, and a little ſpare Pump we had for other Uſes, of three Inches Diameter, worked by one Man, we eaſily maſtered what Leakage we had, and pumped the Water low enough (without ſtaying till the loweſt Ebb of the Tide) for the Maſons to ſet and cramp the Stones of the ſucceeding Courſes: And notwithſtanding this inconſiderable Pumping occaſioned many falſe Reports, and idle Stories, eſpecially among thoſe who never ſaw Works erected in Water, I appeal to thoſe who have ſeen ſuch Works, whether they ever ſaw or heard of ſo few Pumps, or ſo few Hands employed in draining the

the Water over ſo large a Surface as that we had to deal with at a Time.

Laſtly, before the Tide had flown or riſen ſo high, as to endanger the *Caiſſon* and Stone-work from being floated out of its true Place, the Maſons gave over for that Tide, and the Sluice was open'd to let the Water in. As the *Caiſſon* was purpoſely built but 16 Feet high to ſave uſeleſs Expence, the high Tides flowed ſome Feet higher than the Top of its Sides, but without the leaſt Damage or Inconveniency to our Works.

The Maſons afterwards made uſe of all Opportunities, and worked at every Low-water, whether by Day or by Night, notwithſtanding the Coldneſs and Badneſs of the Weather: And the whole broad Pier, and the firſt Courſe of the ſolid Shaft was finiſhed on *March* the 24th.

The Sides of the *Caiſſon* were floated off over the Sides of the Pier on the 30th of *March*, to ſerve the ſame Purpoſes for the *Caiſſon* of the other large Pier, the Bottom of which was launched off from the Tops of the Frames on which it was built, on the 11th Day of *April*; and two Days after, the ſaid Bottom was laid Level upon the Heads of the Stakes or Piles, which ſupported the Frames, in order to receive the Sides which were floated near, and haul'd into their true Place on the 19th Day of *April*, 1739.

After

After the Sides of the *Caiſſon* belonging to the firſt Pier were taken away, the Maſons placed their *Crab* or Engine, with which they hoiſted their Stone, on a temporary Floor, that was faſtened to ſome of the *Fender Piles* which guarded the North Point of this Pier.

The Maſons ſet the laſt Stone of the *Torus* or *Cordon*, on the 20th of *April*; and having cramp'd that laſt Courſe the next Day, on the 23d, being the Feſtival of St. *George*, the *Sheers* and the *Crab* made uſe of in lifting the Stone, and moſt of the neceſſary *Apparatus* for Building was taken away, and this firſt Pier remain'd intirely compleated, having been executed with all the Succeſs that could be deſired, without Loſs of either Life or Limb in any of the Workmen, through the Bleſſing of God; or any Accidents, but ſuch as were eaſily remedied, and attended with a much leſs Expence, than would have attended any other Method of Building the Piers, to the great Mortification of many evil-minded Perſons, eſpecially ſome diſappointed *Projectors* and *Artificers*, who, without knowing what was really intended to be done, or being capable of putting it in execution, roundly aſſerted every where, '*That this* '*Method of Building was entirely impracticable,* '*or at leaſt would prove ſo expenſive, that the* '*Charge of laying the Foundation of one ſingle* '*Pier, would Amount to more than the whole* '*Expence of the Superſtructure.*' But how haſty in their Judgments, and how much deceiv'd in in their Conjectures, these *mistaken Pretenders* have been, the small Expences incurr'd, and the Success we have met with have fully shewn.

I am very far from reflecting on any *Gentleman* having shewn any Diffidence as to the Success of building the Piers as I proposed. Persons that have not been bred Architects or Engineers, and perhaps very little acquainted with Mechanicks, and Hydrostaticks, might reasonably entertain Doubts and Scruples; but for People that are ignorant of the Theory, and unacquainted with the Practice of these Things, to give their peremptory Judgment, contrary to the Opinion of the Board of Commissioners, and to the Sentiments of able Judges and skillful Wo kmen, is what any modest Man should have been ashamed of. As to the many false Reports and Insinuations, that have been maliciously spread abroad, and all that can proceed from Hearts fill'd with *Envy, Hatred, and Malice and all Uncharitableness*; I leave the *Piers of Westminster Bridge*, to give them *the strongest Reproofs*, and the *most solid Answers*.

I have hitherto endeavoured to explain the Method of laying the Foundation of the first Pier, as intelligibly as I could without Schemes or Figuers. The same Method I followed in the Construction of all the other Piers, because the Nature of the Ground being found a Bed of Gravel every where, required no Alteration.

I

I ſhall now endeavour to comply with another Order of the Honourable Board, *viz.* to ſhew wherein the Method I followed, was preferable to others that have been propoſed, and to anſwer the chief Objections that had been made thereto. In order to deal candidly with my Opponents I ſhall give their Objections *their full Force*.

Firſt Objection. *Why ſhould new Methods be employed in the laying the Foundation of the Piers, inſtead of making Uſe of thoſe already practiſed on the like Occaſions?*

Anſwer. The Method I have made uſe of is not abſolutely new: Something like it is often practiſed in the Conſtruction of *Moles*, *Dikes*, *Ramparts*, and other Works erected in the Sea, or in Lakes, or in deep Fens or Moraſſes, and I have aſſiſted in making uſe of it, where all other Methods had proved ineffectual; the Method I mean, is known in *Engliſh* by the Name of *Cheſt-work*, but the Method I have employed differs from *Cheſt-work* in many eſſential Particulars, of which I ſhall only mention two or three. The largeſt Cheſt that I have ſeen or heard of, ſeldom exceeded 16 Feet Square, whereas the *Caiſſon* made uſe of, is near 80 Feet from Point to Point, and near 30 Feet in Breadth; which requires very different Precautions and Care to build and manage from what common Cheſts do.

2dly. When *Cheſt-work* is uſed, the moſt common Method is to fill the Cheſts with Stonework as high as they can without ſinking, and then

then ſuffer them to ſink to the Bottom, and bed there as chance will let them: Whereas this *Caiſſon* was contrived to be let down and raiſed up again without any Difficulty, until it ſhould be found to bed itſelf ſufficiently level upon the Surface of the hard Gravel on the Bottom of the Pit, dug in the River for that Purpoſe.

3*dly*, In the Conſtructions of Works laid in Water by the Help of *Cheſt-work*, the Cheſts are ſunk one after another, as near as they can be, but the whole Work has no Bond to keep it together except its own Weight, whereas each of the Piers of this Bridge are built, Courſe after Courſe, with the ſame Bonding, the ſame Cementing and Cramping, as if built upon dry Ground, and every Pier is, as it were, like one ſolid Stone, laid none leſs than 5 Feet, and none more than 14 Feet under the Surface of the Bed of the River.

2*d* Objection. *Why could not the Piers have been as ſecure, if built on the Tops of Piles driven in the River, and ſawn a little lower than low Water?*

Anſwer. It happens very unluckily for the Objectors, that I cannot anſwer this Objection, without detecting a groſs Ignorance in thoſe that propoſed it. The only Argument I ſhall uſe to refute them, ſhall be a Conſequence of the very Argument which they made uſe of in Support of their Aſſertion, *viz.* that *London Bridge* and *Rocheſter Bridge* were certainly built

ſo,

so, and no doubt but *Bristol Bridge*, *Berwick Bridge*, and in short all Bridges, built in Tide-Rivers, and deep Water, were all constructed by much the same Method; which Method is commonly called *building upon Stilts*.

For allowing this to be true, all that can be deduced from it is, that if the Commissioners ever intended to build such a Bridge as *London Bridge*, they might proceed upon the same Principles; but then this new Bridge would be attended with the same Inconveniences as the old one, *viz.* there must be monstrous Piers, and small Openings, with a Necessity of building Sterlings to preserve the Piers, and a dangerous Fall under the Arches, to the no small Obstruction of the Navigation.

Now I will take upon me to say, that the Commissioners did intend no such Bridge: And indeed they could not admit of any such Proportions between the Arches and Piers; for the first Bridge Act obliged them to leave at least 760 Feet free at all Times for the Water Way; which at low Water does not exceed 1000 Feet.

But those that imagined, that the same Method as was made use of at *London Bridge*, would do to support the Piers of such large arches as were intended (I mean Arches of upwards of Fifty Feet Span upon Piers not above one Quarter or one Fifth of that Opening in Breadth) discovered a very great Ignorance of the lateral Pressure, or the *Thrust* of such Arches. As the intended Brevity of this short Account

Account does not permit me to treat of that Matter in this Place as fully as I could, it will be sufficient to inform the Reader, that if the Piers of the intended Bridge, had been *built upon Stilts*, or on the Top of Piles sawn a little lower than low Water Mark, unless their Bulk had taken up at least one Third, or rather one Half the Breadth of the River, the strongest semicircular Arches that could have been built upon such Piers, would have been in manifest Danger of falling the Moment the Centers that supported them while building, should be struck; and consequently, if the Arches were Elliptical, or only Segments of Circles, as some proposed, they would have been in still more Danger of falling, since the lower an Arch is, in Proportion to its Opening, the greater is the *Thrust* it exerts against its Piers: Besides, it cannot be imagined, that the People who design'd and executed *London Bridge*, and other Bridges in the same Taste, could be so stupid as to have chosen narrow Gothic Arches, supported by monstrous Piers, if they had not known or experienced, that their Manner of laying the Foundation would admit of no other Sort of Arches and Piers. All this is founded on Reason and Geometry, and what is dictated by Reason and Geometry is generally confirmed by Experiments, of which an Instance happened not many Years ago, for the Knowldge of which I am obliged to the late General *Wade*, &c. in whose Hands were the Designs taken upon the

the Place. About the Year 1710, ſome *French Engineers* attempted to build a Stone-Bridge over the River *Allier*, at the Town of *Moulins*, in a Province of *France*, called the *Bourbonnois*; their intended Bridge conſiſted of three large elliptical Arches, ſupported by the two Abutments, and by two Piers, whoſe Breadths were each about one Quarter of the Span of the middle Arch, the whole deſigned in a very bold Taſte.

They placed their Stone-Work on the Tops of a very great Number of large Piles, extending in Surface about ſix Feet all round the outſide of the Stone-Work; there Piles were driven about fifteen Feet deep in the Bed of the River, the Tops of them were ſtrongly faſtened together, and reached about five or ſix Feet above the Surface of the Bed of the River. But the Moment they attempted to eaſe the Centers, on which they had turned the Arches, the whole fell into Ruin. What Reaſon could induce the Perſon or Perſons, who had the Direction of the building of that Bridge, to venture laying the Foundation of ſuch Piers and Abutments in ſo perilous a Situation, is unknown to me; the moſt favourable Suppoſition that can be made, is, that either they were perſuaded the Bed of the River was a Soil of too looſe a Contexture to admit the Building or Draining of a *Battardeau* or Coffer-Dam, of which I am going to ſpeak, or that having attempted to conſtruct ſome *Battardeaux*, they had found them ineffectual by the Water oozing in too faſt upon them, or by ſome Springs unluckily ſtarting in their Foundation; which they could neither ſtop, nor maſter. But to return to my Subject.

This Method of *building upon Stilts* having been exploded by the Honourable Board as inſufficient, I ſhall next examine another Method, which was propoſed to come to the Bottom of the River: For as to all Attempts to turn the whole River, or any conſiderable Part of it, as ſome People propoſed, they would have been attended with incredible Difficulties, an immenſe Expence, and at laſt would have proved ineffectual, as will beſt appear by what follows.

3*d* Objection. *Why could not the Foundation of the Piers have been laid by the Help of Coffer-Dams, ſuch as are called by the French,* Battardeaux?

Anſwer. As it may happen, that ſome Readers may not know what is meant by Coffer-Dams or *Battardeaux*, I beg leave to explain it in few Words.

In the Building of the Piers of Bridges, Sluices, and other Works in Water, Engineers and Architects have often Recourſe to this Method, *viz.* To incloſe the Place intended for the Foundation ſo as to keep the ambient Water from coming in, that it may be drained dry, and kept ſo by Pumping or other Engines. Sometimes this Incloſure is ſingle, and ſometimes double, with Clay ramm'd between; ſometimes the Incloſures are made with Piles only, driven cloſe by one another; ſometimes thoſe Piles are notch'd or dove-tail'd one into another; ſomet mes the Piles are groved, and driven at a Diſtance, and Boards let down between them: But let the Incloſure or Incloſures be made in any of the Ways mentioned, or in any other way, the ſole Intent of this Manner of Proceeding, is only to keep the Water from coming into the Foundation, whenever it can be drain'd.

The firſt Inconveniency attending this Method, is, That if the Incloſure be not ſtrong enough, or not ſufficiently prop'd or brac'd in the Inſide, it will not be able o ſupport the Preſſure of the external Water, (eſpecially if it be Water agitated by ſtormy Winds) which, by breaking and burſting in, often deſtroys many Lives, and entirely defeats the Intentions of the Projectors, that have not taken the neceſſary Precautions, of which I could give a great many Inſtances, ſome of which I have been an Eye-witneſs to. But if this Method had no other Inconveuiency, it could eaſily have been remedied in the Execution of the intended Bridge *England*, and *London* eſpecially, abounding with excellent Artificers of all Kinds: But what would have rendered it entirely uſeleſs, or ineffectual, is the Nature of the Ground under the Bed of the *River Thames*; which at the Place where the Bridge is, is every where a Gravel,

Gravel, covered over on the *Surry* Side with a Sort of loomy Sand; all which would suffer the Water to ouze up (notwithstanding the Sides of the *Battardeauu* or Coffer-dams should be perfectly tight) so fast, especially the Gravel, as to put it out of the Power of any Engine or Engines to drain the *Battardeau* or Coffer-dam: Indeed where the Ground under the Foundation is a stiff Clay, or an Earth of a sufficient consistency to hold Water, *Battardeaux* or Coffer-dams, have been used with Success, tho' attended with an immense Expence and Trouble; and what I would have made use of, if I had not foreseen that it would have been in vain to attempt in this Place to come at the Bottom, and much more so, to reach several Feet under the Bed of the River, by any such Means. Those that have seen (or have been concern'd in) Buildings erected in Water, where the Ground is a Gravel, or a loose Clay, or a Sand, well know the inseparable Difficulties that would have arisen, if such Coffer-dams or *Battardeaux* had been attempted on the *Thames*, over-against the *Woolstaple*; where, besides the Agitation of the Water, occasioned by the Winds, the Height of the Water is perpetually increasing, or decreasing from six Feet to about twenty three Feet, perpendicular Height, above the Surface of the Bed; which two Circumstances alone would make it difficult, and very expensive to provide proper Materials, and construct a Coffer-dam sufficiently strong to resist such unequal Preſſures, ſo as to keep out the ambient Water. As to the Ouzing in of the Water thro' the Pores and Interſtices of the Gravel, looſe Clay, or Sand, it may eaſily be ſhewn, that if all the Interſtices in the Bottom of the Foundation of one of the Piers taken together, amount only to a Hole of ſix Inches Square, (which is a Suppoſition much under the Truth) and ſuppoſing the Tide or Height of the Water above the Foundation (as it is at a mean, or at an Average between the higheſt and loweſt) about 15 Feet perpendicular, they would give above 770 Tons *per* Hour; which is more than 70 Men could Pump out, even ſuppoſing them to act always with the ſame Strength as they do at firſt, and to work Day and Night without ceaſing; and more than 150 Men, or 30 Horſes could do, working as they commonly do.

All that I ſhall add to this Article is, that ſome of the Perſons, who propos'd or eſpouſed this Method of making an Incloſure round the intended Pier, with Dovetail'd Piles, and pretended to drain the Water from within, might remember how fruitleſs was the Attempt, or rather Experiment that was made of it in *Hyde-Park* not many Years ago.

The remaining Objections have been levell'd more particularly againſt the Method I propos'd, which I ſhall now conſider:

4th Objection. *Inſtead of boring the Ground (in order try the Nature of the Bed of the Ri-*

ver)

ver) with a ſharp Tool or Drill, other boring Tools ought to have been made uſe of, and ſome of the Ground ought to have been brought up, whenever it was perceived to alter, in order to ſee the better, what the Nature and Colour of it was.

Anſwer. The Intention of boring the Ground under the Bed of the River, was certainly in order to know whether the Nature and Conſiſtence of it (no matter what Colour) would admit of the Foundation of a ſubſtantial Stone Bridge; now making uſe of a ſharp Tool (the Point of which was not unlike that of a Watch-maker's Drill) anſwered all thoſe Purpoſes perfectly; for by the greater or leſs Reſiſtance we found, we eaſily diſcover'd the Nature and Conſiſtence of the Ground it went through: And by the Tremor of the Bar, and the Noiſe it made as it went in, (which was communicated through the Iron Bar, ſo as to be very ſenſible both to the Ear and the Hand) we were very ſure of the Depth of the ſeveral Strata, whether Dirt, Sand, Clay, or Gravel, ſmall or large, looſe or compact; neither did we rely upon the Drill only, till having tried with other Tools, and brought up ſome of the Ground in many Places, over-againſt *New-Palace Yard*, *Derby-Court*, and the *Horſe-Ferry*; we were always confirmed by the Sight, in what Judgment we had made by the hearing and feeling of the Drill. However, to ſatisfy ſome People's Curioſity, I think proper to mention the Refult of what we brought up to be in general as follows:

The Ground near the Surface is various, generally fmall Sand, Gravel, and Dirt mixed with Bits of Coals, and broken Shells; all thefe of a dirty brown or black Colour: When we found Clay, or a Loomy kind of Sand, they were underneath, moftly bluifh or blackifh, but in fome few Places of a dirty reddifh yellow; in fome Places thefe Clays were foft and clammy, in others ftiff and compact; fome would harden in the Air, but moft would dry and crumble into Sand.

The Gravel next the Surface, was for the moft Part fmall and loofe, unlefs in the Channels of the River, and moftly black and dirty, with the Coals and wafhing of the River. The Gravel, at the Depth of four or five Feet, was of different Sizes, very often of a beautiful reddifh Colour, as that ufed for gravel Walks: And in fome Places the Gravel was mixed with larger Stones, which were cemented together by a ftony Subftance, the whole approaching the Colour and Solidity of the worft Sort of thofe Stones, commonly called *Plumb-pudding Stones.* The Learned may draw what Inferences they pleafe from thefe plain Facts.

But with regard to the main Inquiry, *viz.* the admitting of a good Foundation or not; the Colour of the feveral Subftances under the Bed has nothing to do, and their Hardnefs or Softnefs, and Depths, were all determin'd by the Drill

Drill only; for which Reaſons we left off the Uſe of other Inſtruments, finding nothing by the Uſe of them, but Loſs of Time, and Expence.

5*th* Objection. *Since the Ground under the Surface of the River was found various; in ſome Places Gravel, in others Clay, in others looſe Sand and Mud; and that even under thoſe Places where the Bed of the Gravel was found of a conſiderable Thickneſs, there may be bad Ground, which in Time may perhaps give way; the Foundation of none of the Piers ought to have been laid, without Piling the Ground all over, as cloſe as neceſſary, in order to have no Part of the Foundation weaker than another.*

Anſwer. This Objection would be of great Weight, if it was founded upon real Facts; but as ſoon as the Obſervations of experienc'd Engineers and Workmen are conſidered, it will appear very different from what it does at firſt View.

Whenever a heavy Building or Structure is to be built upon a Foundation that is judged dangerous, the beſt Method is certainly to pile it all over, and having cut the Heads of the Piles to a Level cloſe to the Ground, in the Bottom of the Foundation Pit, to lay the Foundation upon thoſe Piles. 'The Grounds which moſt 'require piling, are looſe or quick Sands, ſoft 'Clays, and ſuch as are found under marſhy 'or fenny Places; but ſtiff Clays, or hard Gra- 'vels, and ſame Sorts of Sands, even in thoſe

'Places where they reach, or extend in Depth
'but an inconſiderable Way, are found by Ex-
'perience to want no piling; ſince Buildings of
'a very conſiderable Weight and Preſſure, are
'found to ſtand firm on ſuch Foundations,
'(without the Time, Trouble, and Expence
'neceſſarily attending Piling) no Matter whe-
'ther the Ground underneath be of a firm Con-
'ſiſtence or not; of which it will be ſufficient
'in this Place, to mention a few Examples,
'purpoſely choſen in or near *London*, for the
'Eaſe of inquiring into the Truth of them.

'The Cathedral of St. *Paul*, (which is cer-
'tainly the fineſt, as well as the moſt ponde-
'rous Fabrick in *England*) ſtands upon a Layer
'of Clay or pot Earth, about ſix Feet thick on
'the North Side, and hardly four Feet on the
'South Side, *and nothing under it but dry Sand,*
'*mixed ſometimes unequally, but looſe, ſo that it*
'*would run thro' the Fingers, for above* 40 *Feet*
'*in Depth*; for a further Account, See *Paren-*
'*talia*, or Memoirs of the Family of the *Wrens*,
'page 285. *& ſeq*;

'*Weſtminſter Abbey* ſtands upon a cloſe ſmall
'Sand, and nothing of a firmer conſiſtence is
'to be found underneath, to the Depth of ſeve-
'ral Feet.

'Part of *Greenwich Hoſpital*, *viz.* the up-
'permoſt Pavillion, ſtands firm and well, upon
'a thin *Stratum* or Layer of Gravel, under
'which is a quick Sand.

The

'The *New Treasury* ſtands extremely firm,
'tho' its Foundations are laid upon a thin Layer
'of Gravel about ſix Inches thick, and under
'which is a very bad Ground, and even a quick
'Sand in many Places.

'Now the Ground under the two largeſt
'Piers, and all the adjacent Parts, was found to
'be a hard Gravel to the Depth of about 14
'Feet, and perhaps it is the ſame for many
'Feet more: Whence it may juſtly be con-
'cluded (due Regard being had to the Nature
'of Things, and the Experience obtained in
'like Caſes, from the Examples cited above)
'that piling ſuch a Ground would have been
'an uſeleſs Expence, and a Loſs of Time, be-
'ſides the Danger of ſtarting Springs of Water,
'with which the Bed of the River is inter-
'ſperſed, and breaking the natural Contexture
'of that hard Bed of Gravel.

'As to that Part of the Objection, which re-
'commends Piling, on Account of having the
'Foundation no more apt to ſink or give way
'in one Part than another, it is of no Weight
'in the preſent Caſe; becauſe the Foundation
'of the Piers is not laid one Stone after another,
'upon the bare Ground, but upon a ſtrong
'Grating of Timber, extending above five Feet
'all round each of the Piers. This Grate con-
'ſiſting of two Rows of ſtrong Timber laid a-
'croſs one another, and ſo pinned and planked,
'as not to admit of any Part giving way, unleſs
'the whole ſhould ſink; which ſeems hardly
poſſible,

'possible, considering the large Surface which
'the Pier and its Grating bears upon.

'Thus much I thought necessary to say, in 'respect to the Piers that have been already e'rected, or the Piers which may hereafter be 'erected without Piling the Foundation, where 'what the Ground shall be found as good as 'was found under the middle Piers. But 'in the Places where the good Ground shall be 'found to lie so low under the Bed, as not to 'be come at without too great an Expence and 'Difficulty, and in such Places where no good 'Ground is found within reach (if any such 'there be) the Piling it all over will become a 'very necessary Precaution."

Thus I express'd my Thoughts and Intentions, in *May* 1739, when only three Piers of the Bridge were built, which I thought fit to republish, and to which it is now proper for me to add, that none of the Foundations was piled, nor wanted piling, because having found a firm Bed of Gravel under every Part of the Bed of the River, where the Piers of *Westminster Bridge* are built, and of so considerable a Thickness, that no boring Tool was able to reach into any other Soil, and having been able to dig down to it under every Pier, tho' under one of them the Gravel was not less than 14 Feet lower than the Bottom of the River, I knew how much better a natural Foundation is, than an artificial one, and therefore justly concluded, that piling in any such Ground, would only break and disturb

disturb the firm Contexture of the Bed of Gravel. All the Master Artificers concerned in the Bridge with me, were always of the same Opinion, and so always declared themselves before the Boards, and every where else; for which Reason I never did or would take any Notice of the many Reports to the contrary, which the Commissioners who attended the Boards or the Works, and every Person concerned in them, knew to be all false, and some of them absurd and ridiculous.

6*th* Objection. *The Timber-frames or Gratings, on which the Foundation of the Piers are laid, will rot or decay in Time, and thereby occasion both the Ruin of the Piers and Superstructure; and this is the more to be feared, and the sooner to be expected, as the Timber made use of in the Foundation is only Fir, which will always lie soaked in Water or wet Ground.*

Answer. This Objection, like most of the others, is made by Persons that are not acquainted with the Facts they should know, or might know, if they would give themselves the Trouble of enquiring into Things, or asking Questions of experienced People, who would have told them, that not only *Oak* or *Elm*, but even *Fir*, (provided it be sound when laid down) will never decay, if it be always kept under Water from the Contact of the Air.

That Timber of all these Sorts have been taken up, which were known to have lain wet or or under Water for many hundred Years, the Surface of which, inſtead of being decayed, ſeemed rather to have harden'd, in Proportion to the Time the Timber had lain there; and in ſome Places the Surface, ſuch as the Bark or the Sap, was found putrified, but the Inſide remained ſound, (tho' of a darker Colour) and that the Inſide of ſuch Fir Timber, in particular, was found to have preſerved even its reſinous Particles, ſince it diſcovered a ſtrong Smell of Turpentine upon the firſt Stroke of a Chizel.

To ſay, that the Timber ſo taken up moulders away in the open Air much ſooner than if it had not been ſoaked ſo long, is nothing to the preſent Purpoſe, ſince no Part of the Foundation is intended to be taken up, neither is it at all material (if any Part ſhould hereafter be taken up) how long it would be a rotting or mouldring away, when expoſed to the Air.

Gratings of Timber are put under every Pier, in order to ſtrengthen the Foundation, and mend it, in caſe it was not perfectly of the ſame Degree of Hardneſs every where: And it is a Precaution, which good Engineers often obſerve in the Foundation of *Ramparts*, and other heavy Buildings.

As to the Gratings being Fir, that Timber was choſen as being the cheapeſt, the eaſieſt worked up, equally ſerviceable for the Purpoſe, and as often employed as any other; what occaſions Timber to decay moſt, is, its being ſome-

ſometimes wet and ſometimes dry; eſpecially if the Timber was not cut in a proper Time, or properly ſeaſoned when employ'd. It is for this Reaſon, that the Piles that ſupport Timber Bridges, (particularly in Tide Rivers) and the Timbers employ'd in Wharfs and Cauſeways are ſo ſoon decay'd, becauſe in thoſe Places they are often wet and dry; tho' at the ſame Time the Parts that are always above the Water, do not decay ſooner than in other Places, and the Parts under Water always remain ſound and good.

7*th* Objection. *Notwithſtanding all the Precautions that have been taken to render the Foundations of the Piers as firm and ſolid as if they were built upon dry Ground; they will always be in Danger of the Water gulling underneath, and carrying away the Ground from under the planked Gratings on which the Piers ſtand.*

Anſwer. This Objection is general againſt the Duration of all Buildings erected in running Water: but the Moment we conſider Facts, and apply Numbers, it will appear of no Weight in the Caſe of the *River Thames.*

The Effect produced by the Gulling of a River (every thing elſe being ſuppoſed alike) is nearly in Proportion to the Velocity of the Stream, which is confirmed by Experience; for, in all ſlow running Rivers, the Gulling is ſo inconſiderable, that the Stream is not able to carry away the Mud, and other light Subſtances: And to mention the op-

poſite

posite Case, in swift running Rivers or Torrents, not only Mud, Sand and Gravel, but large Stones and Trees are carried away according to the Rapidity of the Stream. Now it is evident, that the Velocity of the Stream under this Bridge, does not so much as equal the Velocity it has at the *Horse Ferry*, or just *above London Bridge*; at either of which Places, the *Thames* may very justly be called a gentle River, since it is not able to gull or carry away any Part of the Gravel, and therefore it is not at all likely that the Water should gull away the Ground under the Foundation of the Piers, considering what an immense Weight presses upon them.

I know it has been urged, in order to defend this last Objection, that if *London* Bridge is ever mended, by lessening the Sterlings, or by taking some of the Piers away, or entirely rebuilt, the ebbing Waters will fall to the Sea in a less Time than they do now; and, consequently, that the Velocity of the Stream under this Bridge will then be increased, and occasion the Ruin of the Piers, on Account of the Gulling of the Water under them.

Now, to shew how idle Objections are, when proposed only in general Terms, I will suppose the worst, *viz.* That *London Bridge* should be entirely taken away; the utmost that could happen then, would be that the ebb Tide would

run

run much leſs Time than it does now: Let us again take the worſt, and make it of no greater Duration than the Tide of Flood, that is, a little more than ſix Hours each; it will follow in that Caſe, that all the Water that now takes up about eight Hours in Ebbing out, would Ebb out in about ſix Hours; and, conſequently, the Velocity of the Water would be increaſed nearly in the Proportion of 6 to 8. or of 3 to 4, from what it is ſince the Bridge is built.

But ſtill the Stream would be very gentle, and (to quote a near Example) would be gentler under the Arches of this Bridge, than the Stream of the River *Seine* is under the Arches of the *Pont Neuf*, or *Pont Royal* at *Paris*, where that River is experienced not rapid enough to occaſion any Damage to the Piers of thoſe Bridges, by gulling under their Foundation.

I rather ſuſpect (with regard to the *River Thames*, and *Weſtminſter Bridge*) the Reverſe of the Objection will happen, *viz.* That the Piers will, in Time, be more and more buried in the Ground, by the Silting of the River; which muſt accumulate in a long Courſe of Years; for in all Tide Rivers (at leaſt thoſe that have no extraordinary Declivity towards the Sea) we find their Bed continually riſing; and in the *Thames*, in particular, it is well known, that the Bed of the River (above *London Bridge* eſpecially) is ſeveral Inches higher than it was 100 Years ago; which I conceive to be occaſioned

2 by

by the Tide of Ebb, having ſo long a Time to depoſite its Settlement, and every Tide of Flood hindering in a great Meaſure part of that Settlement, from being carried down to the Sea.

APPENDIX.

APPENDIX.

CONTAINING

Some Particulars relating to *Weſtminſter Bridge*, or to the Hiſtory of the Building thereof.

AT the earneſt Deſire of ſeveral Perſons, I have annexed the following Particulars, for the Sake of thoſe who may deſire to be ſatisfied in ſeveral Things, without waiting for the Publication of a larger Work on the ſame Subject, which I have undertaken, and do deſign to compoſe and publiſh as faſt and as ſoon as my Health will permit, and of which I have hereunto added a fuller Account, with the Table of Contents.

In the firſt Place, I think proper to inform the Readers, that *Weſtminſter Bridge*, and the new Streets and Roads leading to and from

 the

the ſame have been laid out, made and built by a Commiſſion appointed and continu'd by the Authority of ſeveral Acts of Parliament, conſiſting of near 200 Lords and Commoners, all of them Members of one or the other Houſe of Parliament at the Time of their Nomination, who notwithſtanding their great Trouble, Care, and weariſome Attendance, in the Diſcharge of the ſeveral important Truſts repoſed in them by the Legiſlature, have abſolutely no kind of Salaries, Perquiſites, Fees, Rewards or Conſiderations whatſoever, except (as a Nobleman among them nobly expreſſes it) *the Honour of doing what was thought impoſſible.*

Next to this, I will mention, that though the Variety of Affairs committed to the Care of the Commiſſioners obliged them to employ ſeveral Perſons as Officers under them, *two and no more*, were employed ſolely and directly in the Building of the Bridge. Their reſpective Provinces were, abſolutely independent the one from the other, the Deſigning and Conducting the Works, and the Directing the Perſons concerned therein, was committed to one Perſon only, under the Name of *Engineer*, and he was exempted from having any Thing to do with any Money Affairs, or any thing relating to *Contracts or Agreements*, *Quantities*, *Qualities, and Prices of Materials*, *and Workmanſhip*, *Meaſurements*, *Accounts of Days Work and Extra Work*, *Allowances for Waſte*, *Damage*, *Salvage*, *and contingent Charges*; the keeping proper Accounts

counts of these was made a separate Province, and allotted to *Richard Graham*, Esq; under the Name of *Surveyor and Comptroller of the Works*, (the first Time I believe, that both those Appellations were given to one Person) and after his Death, (which happened in *May* 1749) to Mr. *Obadiah Wylde*, whom the Commissioner, appointed their Clerk of the Cheque, towards the Conclusion of the Works. In large publick Works, such as this, Clerks are commonly allowed to the two first mentioned Officers, but the Commissioners had so great a Regard to Frugality in the disposing of publick Money as not to allow of any such Helps. The deceased *Mr. Graham* did indeed employ a Clerk, which he paid out of his Salary, but as I was always both able and willing to do my Duty in Person, and not by Proxy, I never employed any Clerks or Assistants, nor ever wanted, or asked for any.

Above nine Parts in ten of the Works (and in short all that could be done by Contract) was contracted for by the Commissioners, who never entered into those Contracts, but with the greatest Caution, obliging the Persons they contracted with, to find all such Securities as were directed by the Acts of Parliament, and limiting them as to *Time* as well as to *Prices*, which Contracts being once passed, it was not in the Power of all the Officers of the Commissioners put together (were it ever so much their Interefts or their Defires) to retard the Execution of the Works, one fingle Day.

All Salaries were regularly paid quarterly, as they became due, all Bills of Days Works, Materials delivered, or Payments on the Contracts (after having been duly examined and certified by the proper Officers) were referred to the Commiffioners Committee of Accounts, by whom they were re examined, ftated, audited, and when approved, ordered to be paid by Warrants under the Hands of a fufficient Number of the Commiffioners prefent at a Board, and directed to their Treafurer, who always paid the whole full Amounts of the faid Warrants, without any Fee or Deduction whatfoever.

The laft Particular which I think worth mentioning as to Management, is that no Fee or Perquifite, of any Sort whatfoever, was ever allowed to any of the Officers or Perfons employed by the Commiffioners: I am fure I can very truly affert, that I never received, demanded, or even expected any, and I believe every one of the Officers may fay the fame.

The following Particulars relate to Time and Dates.

IN the Summer Months of the Year 1734, feveral public fpirited Gentlemen had Meetings to fet on foot a new Application for obtaining a Bridge at *Weftminfter:* They made a Contri-

Contribution among themſelves for that Purpoſe, and conſulted their Repreſentatives in Parliament, who approved of the Deſign, and promiſed their Aſſiſtance ata proper Seaſon; they alſo procured from me and others, the neceſſary Maps and Surveys, cauſed the Bed of the River to be meaſured, ſounded, and bored at ſundry Places, with Obſervations, and Calculations, *&c.* A Petition to the Houſe of Commons was by them intended to be preſented in the Spring following, but as it could not be got ready and ſigned in Time, it was not preſented before *February* the 4th 1735-6; the Purport was to have a *Bridge erected at the Horſe-ferry or at ſuch other Place as the Houſe ſhould think fit.*

The Houſe of Commons paſſed a Bill that Seſſions accordingly on a Diviſion, 104 againſt 12. The Lords paſſed it with Amendments, to which the Commons agreed after Debates, and the Royal Aſſent was given on *May* the 20th 1736.

The Commiſſioners appointed by this Act were near 200 either Lords or Members of the Houſe of Commons. The Bridge was (by this firſt Act) to be erected *from New Palace-yard to the oppoſite Shore*; and there was granted 100,000 *l.* towards it, to be raiſed by a Lottery.

This firſt Lottery proving unſucceſsful, a new Lottery was granted by a ſecond Act in 1737, by which Leave was alſo given to build the Bridge at the *Horſe-Ferry, or at any other*

Place in the Pariſh of St. Margaret or St. John, Weſtminſter.

On the 6th of *July* 1737, a Deſign for a Bridge at *Palace-yard* was preſented to the Commiſſioners, which conſiſted of ſeventeen Arches, the Breadth of which was to be but thirty Feet, the Outſide was propoſed to be *Portland* Stone, the Inſide, Chalk and Rubble, and it was to be *built upon Stilts*; for all which the Sum of 124000*l.* was demanded, but this Propoſal was rejected. At the next Board, the Commiſſioners reſolved that the Bridge ſhould be *intirely of Timber*, and erected at the *Horſe-ferry*, on a Diviſion which appeared at firſt thirteen againſt thirteen, which was determined as juſt now mentioned, by the Chairman's caſting Vote. But the real Numbers were fourteen to twelve, their Clerk (by miſtake) having been told for a Commiſſioner; they then ordered Plans and Propoſals to be laid before them accordingly, on the 29th *July* following, at which Reſolutions the Publick both in Town and Country ſhew'd great Diſſatisfaction.

On the 24th of *Auguſt* following, inſtead of approving of any of the Deſigns offered for a Timber-Br dge, or contracting for any .with Workmen, they reſolved (after Debates) that it was moſt adviſeable to build two Piers at their own Charges, and at their ſucceeding Meetings, they conſidered of ſeveral Methods then offered to them, to build the ſaid Piers, none

none of which did appear to them either proper, or practicable.

On the 31st of *August* 1737, I received a Letter from Sir *Joseph Ayloffe*, Bart. Secretary to the Commissioners, containing a Copy of the following Order.

Bridge Office, *Duke-Street*, *Westminster*, the 31st of *August* 1737.

Ordered, *That Mr.* Charles Labelye *be desired to produce at the next Board, his Model for explaining his Method of laying the Foundations of the Stone-Piers of a Bridge, below the Surface of the Bed of the River.*

By Order,

Jos. Ayloffe, Cl.

On the 7th of *September* following, I attended on the Commissioners accordingly, and had the Honour of explaining before them, upon a working Model, the Method I proposed to be followed in the Building of the Piers of the (then intended) Bridge, of which they took so far Notice as to come to the following Resolution.

Resolved, *That this Board do approve of Mr.* Labelye's *Design, and are of Opinion that he is a proper Person to be employ'd, in case the Commissioners proceed to the laying the Foundation of Stone-Piers.* And then they adjourned to the 1ſt *Wedneſday* in *February* following, *viz.* 1737-8.

In the Spring, the Commiſſioners obtained a third Act of Parliament, in which (to avoid all future Diſputes) the Place where *Weſtminſter* Bridge was to be built, was fixed at or near the *Woolſtaple*, a little lower than *New Palace-Yard*, but the Choice of the Deſign and the Materipls were ſtill left to the Commiſſioners, and their Powers were not only confirmed but enlarged.

In the ſaid Month of *February* 1737-8, at the Commiſſioners Deſire, and before a very large Board, I had the Honour of explaining a ſecond Time my Method of building the Piers, which appeared to them the cheapeſt and moſt practicable. They reſolved not long after, that the Bridge ſhould ſtand upon Stone Piers, and fixed their Number and Dimenſions according to my Deſign for a Stone-Bridge; they alſo did me the Honour to appoint me to have the Direction of the Execution of the ſaid Piers, under the Name of *Engineer*---and they were pleaſed (at my earneſt Requeſt) to exempt me from meaſuring the Works or keeping any Accounts of the Materials delivered or Work done, which was made the Province of the *Comptroler and Surveyor*, both our Commiſſions bearing Date the 19th of *May* 1738.

The Commiſſioners reſolved alſo about that Time, that the Superſtructure of the (then intended) Bridge ſhould be of *Oak Wood*, according

ing to a Deſign of the late Mr. *James King*, with whom and his Par ner Mr. *John Barnard*, they contracted for ſuch a Supe ſtructure, for the Sum of 28000 *l.* the whole to be compleatly finiſhed within twelve Calendar Months after the finiſhing of all the Piers.

On the 2d of *June* 1738, the Commiſſioners iſſued out the firſt Orders to proceed to the Execution, which was immediately taken in Hand, by preparing Piles and other Timbers, and making the neceſſary Engines in a ſmall Spot of Ground on the *Surry Shore*.

The 13th of *September* 1738, we began working in the River, by driving the firſt Guard Pile for the Incloſure of the Weſtern middle Pier.

On the 29th of *January* 1738-9, the firſt Stone of *Weſtminſter Bridge* was laid by the *Late Right Hon. the Earl of* Pembroke, *&c.*

On *April* the 23d 1739, the firſt Pier was finiſhed.

On the 26th of *December* following, the *Thames* begun to carry Ice, and the Froſt increaſing, a total Stop was put to the Works till the 18th of *February* following, and the Damages not made good till the 19th of *March* following. The Ice carried off all the Piles then ſtanding, about 140 in Number, and broke above half of them, as well as a Sett of Sides of one of the *Caiſſons*. At that Time three of the Piers were built, the *Weſtminſter Abut-*

ment

ment was brought above low Water Mark, and the Foundation of the *Surry Abutment* was begun.

On *January* 31ſt, 1739-40, on a Motion made by *the Right Honourable the preſent Earl of Bath*, I was ordered by the Board to lay before them, on the 13th of *February* following, a general Eſtimate for a Stone Superſtructure; Summons were ſent accordingly to 184 Commiſſioners, out of which 36 attended (which was the greateſt Number preſent at any one Board, either before or after) on that Day, after Debates, (*the Right Honourable the Preſent Lord Viſcount Falconberg in the Chair*), it was reſolved by a Majority of 20 to 12, that the Superſtructure ſhould be Stone; and I was ordered to prepare and to lay a Deſign for the ſame before the Board on the 5th of *March* next.

Soon after the Board agreed with their Carpenters, that they ſhould deliver up all the Timber already provided by them; that they ſhould be repaid all their D ſburſements, and be allowed Eight *per Cent.* (all they aſked) on their whole Contract, *viz.* 28000 *l.* which upon theſe Conditions they gave up.

On the 5th of *March* I preſented a new Deſign, which was very near the ſame as the Bridge is now finiſhed; but the Board not being very numerous, they adjourned to the 12th next, on which Day my new Deſign was approved as high as the Top of the Balluſtrade, and ordered directly into Execution.

On

On the 5th of *May* 1740. was begun the fourth Pier, which was finiſhed in twenty Days. *N. B.* I thought this fact worth mentioning, to ſhew the Readers what Diſpatch was made by this Manner of building, whenever a ſufficient Quantity of *Portland* Stone could be procured from the Quarries.

On the 14th of *Auguſt*, 1740. was begun the Building of the Arches, by ſetting the N. W. Springers of the middle Arch, One Year, Six Months, and 16 Days after the laying the firſt Stone of this Bridge.

On the 21ſt of *February* 1743-4. the 14th and laſt Pier was finiſh'd; ſo that the building of both the Abutments, and all the Piers of this Bridge were happily compleated in five Years, and twenty-three Days, from the laying the firſt Stone, notwithſtanding all the Stops and Difficulties occaſioned by the Tides, bad Weather, Ice, and frequent Wants of Stone, which was kept from us by long eaſterly Winds, beſides ſome Embargoes, extraordinary Preſſing of Seamen, and ſtaying often for Convoy in Time of War.

On the 20th of *July* 1746. the laſt Arch was key'd in, five Years, eleven Months, and five Days after they were begun, and from that Day (if it had been proper) the Bridge was paſſable both for Foot Paſſengers and Horſes, which was compleated in ſeven Years, five Months, and twenty-two Days, from the laying the firſt Stone.

On

On the 25th of *October* 1746. the laſt Stone of all the Abutments, Piers, and Arches was laid, (but without any Ceremony) by the *Late Right Honourable the Earl of Pembroke*, who had laid the firſt but ſeven Years, eight Months, and twenty-ſeven Days before, and excepting the Foot Pavements and Balluſtrades, the Bridge was then intirely finiſhed and fit for Service.

On the 25th of *July* 1747. the laſt Center was taken down, and all the fifteen Arches of this Bridge were left intirely free and open.

On the 14th of *November* 1747 the Bridge, and the Roads and Streets on both Sides, were compleatly finiſh'd, and the whole was performed in ſeven Years, nine Months, and ſixteen Days, from the laying of the firſt Stone.

The Commiſſioners intended ſoon after this to have opened the Bridge for the Service of the Publick, but were prevented by an Accident intirely unforeſeen, and not eaſily accountable, of which I ſhall give but a very ſhort Account in this Place, my intended Brevity not allowing me to enlarge thereon, as I ſhall do in the laſt Chapter of the firſt Part of the ſecond Volume of a Work which I have undertaken to publiſh hereafter, the Contents whereof the Reader will find annexed at the End.

In the Months of *May* and *June* 1747, the Weſtern fifteen Foot Pier of *Weſtminſter-Bridge* was perceived to ſettle very gently at firſt, but ſo much faſter towards the End of *July*, 1747. that it was thought abſolutely neceſſary to take

off

off the Ballustrades, Paving, and Part of the Ballast, that laid over the said Pier, and the two Arches adjoining; by the Continuation of the settling of this Pier, those Arches lost their regular semicircular Figure, considerable Openings in the Joints shewed those Arches in some Danger, and some of their Stones both in the Fronts and their Sopheits were split and broken, one of them actually fell out of the least Arch in the River, and another was taken out to prevent its falling down.

Notwithstanding most of the considerable Bridges of which we have any Account, have in the Course of their buiding, met with some Accident like this, it is certain that never was such an Accident so much taken Notice of; it was very sincerely deplored by all those who had any good Nature, or publick Spirit, and as heartily rejoiced at by those of a contrary Disposition, such as the Watermen, Ferrymen, *&c.* and great many others, nay by some, who were fed and maintained by the Commissioners, with much better Bread, than ever they deserved, or ever could earn.

In *August* and *September* 1747. many of the Commissioners returned to Town from their Country Seats on this melancholy Occasion, reassumed their Boards, had frequent Meetings, and asked the Opinion of every Person whom they thought able and willing to assist them with their Advice; they received Schemes and Proposals from Abundance of People, most of them

absurd,

abſurd, improper, or impracticable, but in the great Number, ſome there were far from deſpicable, though faulty, for Want of the Projectors being truly informed of the State of the Caſe. Till the Commiſſioners could fix on the moſt proper Remedy, and the moſt adviſeable Way of Proceeding they previouſly reſolved, according to my Advice, in which both the Maſter Maſons, Meſſieurs *Jelſe* and *Tufnel*, unanimouſly joined, that the firſt neceſſary Steps were to build the two Piers next to the two Arches and Pier that were damaged, quite ſolid in Rubble Stone, and Mortar, and to load them ſufficiently, in order to preſerve all the other Arches and Piers of the Bridge; which was ordered accordingly, and immediately after ſet about and done.

The next Step reſolved upon by the Board, on the unanimous Opinion of all the Perſons concerned in the Works, was, that two Centers ſhould be immediately framed, and ſet up as ſoon as poſſible under the two damaged Arches, in order to ſave them from falling, in caſe the fifteenth Foot Pier ſhould ſettle much more, and to afford a ſafe Way of taking them down, if it was found neceſſary to be done. Theſe two Centers were ordered into Execution as early as the 15th *September* 1747. All the former Centers had been contracted for by the Foot, including the Iron Work, the Setting them up, and all Workmanſhip to be performed at a fixed Price, and within a certain limited Time; but the Price of the

Timber

Timber only contained in theſe being agreed for and all the Workmanſhip, and the Setting them up being ſuffered to be done by Day-Work; the ill Conſequence of this was, that notwithſtanding my moſt earneſt Sollicitations, and repeated Orders to uſe Diſpatch, from the Commiſſioners, the ſaid two Centers coſt great Sums of Money, and were above ſeven Months Time in making and ſetting up; during all which Time nothing could be down towards repairing the damaged Pier and Arches.

On 12 *April* 1748. the Maſter Maſons joined with me in adviſing the Commiſſioners to load the Pier that had ſettled, as much as poſſible, which plainly appear'd to us to be the beſt Way to try it, and to ſecure it for the future. Orders were given accordingly, and ſoon after we begun loading the ſaid Pier with Blocks of Moor, Stone, and large Iron Guns, cóndemned as unſerviceable (obtained from the Office of Ordnance) and with Rubble in their Interſtices, by which Means we depreſſed the Pier a few Inches lower: The whole Weight of the Load placed on the ſaid Pier was ſo far magnified by Writers of daily News, and monthly Magazines as to be called 12000 Tons, but it did never exceed 700 Tons, which was but about the third Part of what I intended to load it with, (and what could have been done at a very little Charge, and in a ſhort Time) if my Opinion and that of the Maſons had been followed. The Reaſon why it was not followed, was this; ſoon after the Accident

cident of the ſettling of this Pier had happened, I perceived (but, I own, not ſo ſoon as others did) by the Looks, and even the open Exultations of ſome, and the dilatory Proceedings of others, that a *wicked Cabal*, bent upon Miſchief, chiefly for Miſchief Sake, was reſolved to exert themſelves to their utmoſt, in order to diſtreſs and obſtruct the Works of the Bridge, and if poſſible to hinder its being ever finiſhed as was intended. I could (if I thought it proper) paint the odious Characters of every Fool or Knave of which this Cabal conſiſted in their true Colours; but as this would be no Way intereſting to the Readers, I ſhall proceed by informing them, that this *Cabal* had Credit enough (at that Time) with ſome of the moſt active among the acting Commiſſioners, to make them believe that the further Loading of the ſettled Pier would be dangerous to the Arches, which, they ſaid, might thereby cruſh the Centers, and fall in the River, and even draw after them what they called *a conſiderable Part of his Majeſty's Ordnance*; as I knew theſe Aſſertions to be falſe and groundleſs, I called to my Aſſiſtance four ſkilful and experienced Builders and Maſter-Carpenters, who after a full Enquiry, and a careful Survey, joined the Maſons and me in Opinion, and gave it under their Hands, *that the two Centers were in no Danger, and might, and ought to be eaſed, and the loading on the Pier increaſed as much as poſſible*; but ſome of the Commiſſioners being really terrified at theſe imaginary

imaginary Dangers, moved the Board to ceaſe loading, and even to unload the ſaid Pier, which Motion (after Debates) they carried at laſt by a Majority of two Voices only (if I am not miſinformed) and in Conſequence, on the 5th of *July* 1748, I received Orders to unload the Pier, and to proceed next to take down the two damaged Arches. This Order was the firſt and the only one that ever I received from the Commiſſioners, contrary to my Judgment or Opinion, and which I obey'd, *but I own not without ſome Concern.*

Whilſt the Arches were unbuilding and taking down, the Commiſſioners continued receiving Propoſals and Schemes for repairing the ſaid damaged Pier and Arches, and examined ſuch of them as would bear any Examination.

On *September* the 27th 1748, they ordered me to explain, at large, and to lay before them what Schemes I had to propoſe for the ſame Purpoſe, in Writing. My Report was thoroughly examined on the 13th of *December* following, and (after I had explained wherein all the beſt of the Schemes propoſed were deficient or impracticable) of the two Schemes I propoſed, as equally practicable, that which the Maſons and I ſeemed to incline moſt to (on Account of its preſerving the Regularity and Symmetry of the Bridge, and it being the cheapeſt of the two, the eaſieſt in the Execution, and laſtly, requiring the leaſt Time) was

G approved

approved by the Commiſſioners, and ordered into Execution (under my Direction) as ſoon as the Seaſon would permit, to the great Diſappointment and Mortification of the whole Cabal, and of a great many other Projectors.

It is not poſſible to explain, in a very clear Manner, the Method which I then propoſed, and have ſince put in Practice, with Succeſs, for the repairing the damaged Pier and Arches, without annexing ſeveral Copper-Plates which is not ſuitable to this ſhort Tract: However, to give the Reader ſome Idea thereof, I ſhall only ſay, that the firſt Steps, after the Arches were taken down, was to take down alſo a conſiderable Part of the ſettled Pier, which was at firſt built ſolid a good Way up between the Arches.

The next Step was to encloſe, by a Caſe of large dove-tail'd Piles, the Gravel (in which this and all the other Piers were built) and whatever other Soil there might be under the *Straum* or layer of Gravel, from giving Way, ſpreading, or ſlipping from under this Foundation. Theſe dove-tail'd Piles were driven all round cloſe to the Bed of Timber, on which the Pier is built, and ſo deep as to reach about fifteen Feet under it all round, and afterwards were all ſawn low enough below low Water Mark, as never to be any Obſtruction to the Navigation of any Boat or other Veſſel.

Then the two damaged Arches were rebuilt the very ſame in Appearance, but with much leſs Materials in the Inſide, which I contrived chiefly

chiefly by Means of *a Counter-Arch*, over the settled Pier, between the two Arches and two *Semicounter Arches* butting against the opposite Side of the sixteen and fourteen Feet Piers. So that upon the whole, the Weight pressing on the Foundation of the Western fifteen Foot Pier was lessened in the Proportion of nearly four to three; and as it was known what Weight the Pier bore when it ceased settling, so it may be confidently asserted, that the Present Western fifteen Foot Pier (though in all Appearance the same as it was at first) supports several Hundred Tuns less Weight than it did before.

The next Winter, and Part of the ensuing Spring 1749, were wholly employed in taking down the damaged Arches and Turretts.

About the End of *April* 1749, by the Removal of one Person, and the Death of another (which happened soon after) the Cabal was totally routed, its mischievous Intentions seen through, and prevented, and Peace and Harmony returned again among all the Persons concerned in the Works of the Bridge. The Rooms of those removed or dead, being filled up with Persons fully as honest, at least, more capable, and much more willing to do their Duties, the Repairs wanting were then pursued with Alacrity, and a Dispatch which made itself remarkable not only to the Commissioners but to the Eyes of the Publick.

On the 26th of *April* 1748, the Commissioners appointed for their Master Carpenter

 Mr

Mr. *Edward Rubie*, whom I recommended to them, and had known many Years for a Man of ſtrict Probity, and of great Skill and Experience in all Works of this Kind; and he has ſince very diligently and fully anſwered all Expectations.

On the 17th of *July* following we began driving the dove-tail'd Piles, and by an Accident which happened to the Engine the next Day, were prevented from driving any more till *Auguſt* the 3d following, when we begun driving again without any more Interruption, every Tide, almoſt Day and Night, by which Diligence we had them all driven by the 5th, and all ſawn off by the 20th of *December* following, at which the next Board expreſs'd ſo much Satisfaction, that they ordered a particular Minute thereof to be inſerted in their Minute Book. The three following Months were employed by the Maſons, in reſtoring the Pier (that had ſettled a little out of an upright) to a perpendicular Poſition, which being done by chipping only, took a good deal of Time, and by the Carpenters in new framing and ſetting up the Centers, which, together with a hoiſting Stage, were both compleated the 18th of *April*, 1750.

The Maſons begun the re-building the two Arches on the 23d of *April*; and tho' by their Contract, they were allowed 12 Months Time to finiſh them in, after the Centers and hoiſting Stage were compleated for their Uſe, the Materials being all at Hand, they uſed ſuch Diligence,

gence, that both the Arches were keyed by the 20th of *July*; ſo that there really was much leſs Time employed in rebuilding them than in taking them down. The Centers were eaſed and ſtruck from under them on the 27th and 28th of the ſame Month, and neither of the new Arches deſcended or followed their Center ſenſibly, not even ſo much as the Thickneſs of a common Packthread; the three following Months were employed in building the Counter Arches, filling the Spandrels, building the Foot-ways and Balluſtrades, and in clearing the Bridge, and gravelling the Carriage-way the whole Length of it; and after working at laſt for a few Weeks, Day and Night, *Weſtminſter Bridge* was compleatly finiſhed on *Sunday* the 18th of *November*, 1750. Eleven Years, Nine Months, and Twenty-one Days, after the laying the firſt Stone, which was on the 29th of *January*, 1738-9.

I conclude this Article by mentioning, that on the ſaid *Sunday*, the 18th of *November* 1750, *Weſtminſter Bridge* was opened for the Service of the Publick, by Order of the Commiſſioners, of which previous Notice had been given for ſeveral Days before, in the *London Gazette*, and in ſeveral other publick News Papers.

The

The following Particulars are inserted, to give the Readers some Ideas of the Size and Magnitude of the Parts of Westminster Bridge, *and of the Quantity of Materials employed therein.*

FOR the Sake of some Readers I shall mention, first, That what is meant by a *Ton* of Stone, is the Quantity of about 16 cubical Feet, which weighs generally 20 Hundreds, each Hundred Weight containing 112 *lb. Averdupoize*; and that 40 cubical Feet of Timber is called a Load of Timber, which (if dry Oak) weighs much about the same as a *Ton* of Stone.

Three intire *Caissons* were built to lay the Foundations of the Piers, with all possible Dispatch, each of which contained above 150 Load of Fir Timber, and was of more *Tonnage* or Capacity than a Man of War 40 of Guns.

As to the Size of the Piers and Arches, the Ground Plan of one of the Middle Piers having been laid down in Chalk Lines, on the Floor of the *Painted Chamber*, near the *House of Lords*, there was hardly Room to go along the Walls without treading upon the outward Lines.

Lines, though that Room is full 66 Feet by 25; ſo that the Inſide of one of the *Caiſſons* in which the Piers were built, is, as to Length and Breadth, very near the ſame as the ſaid *Painted Chamber*. And the Floor of the ſaid *Painted Chamber* was found, upon Trial, not ſpacious enough to lay down upon it one half of the Front of the Middle Arch.

The Inſide of *Weſtminſter Hall* having been carefully meaſured, in *June*, 1742. was found 239 Feet 5 Inches long, 66 Feet 10 Inches broad, and 65 Feet 6 Inches high, from the Pavement to the Crown of the *Wooden Truſſes*, or rather *Arches* under its Roof; whence it follows, that ſince the Middle Arch of *Weſtminſter Bridge* is 76 Feet wide, each of the two adjoining Arches 72 Feet wide, and each of the two next adjoining theſe laſt on each Side, 68 Feet wide, there are therefore five Arches in *Weſtminſter Bridge*, each of which is wider than the Inſide of *Weſtminſter Hall*, *viz.* the Middle Arch wider by 9 Feet 2 Inches, each of the two adjoining Arches wider by 5 Feet 2 Inches, and each of the two next adjoining Arches wider by 14 Inches only.

As to Quantity of Materials, it can be proved, that the two Middle Piers of *Weſtminſter Bridge* contain full 3000 cubic Feet, or near 200 *Ton* Weight of ſolid *Portland* Stone, more than is contained in all the Stone employed in the *New Treaſury*, not only in the Fronts, but in all the

 Inſide,

Inside, Stair-cases, half Paces, and even Cisterns included.

It can also be proved, that the Quantity of solid Stone contained in the Middle Arch only, above its Piers, exclusive of the Freese, Cornish, Plinths, Ballustrades, and Foot-ways, is full 500 *Ton* more than double of the whole Quantity of Stone in the *Banquetting House* at *Whitehall.*

Another Particular worth mentioning, is, that there is above fifty thousand Pounds worth of Stone and other Works, in the Piers and Abutments of *Westminster Bridge*, always under Ground or under Water, besides what can ever appear above the lowest Low-water Mark.

And lastly, that the Quantity of Stone Materials in *Westminster Bridge*, is near double the Quantity of the same Materials in the *Cathedral of* St. *Paul*, *London.*

The following Particulars relate to the Expence that has attended the Building of Westminster Bridge.

AS I was exempted, at my own Request, from having any Concern with Measurements, or any Money Affairs, the Readers, I hope, will excuse my not being able to give them so full Satisfaction upon this Article, as they might desire. I am informed, that an able Person

Perſon has kept an exact Account of all the Receipts and Expenditures of this Commiſſion, reduced under proper Heads, of which the Bridge itſelf is one, and certainly the moſt conſiderable; but I am intirely ignorant when or whether that Account will ever be publiſhed. In the mean Time, before I mention what (as I am inform'd) is the Total Amount of the neat Expence of this Bridge, I believe the Readers will not be diſpleaſed, to hear of ſome of the Means that have been uſed to co-operate with the remarkable Frugality which the Commiſſioners have ſhewn all along in the laying out ſuch Sums of publick Money, as they have been intruſted with from Time to Time, by the Legiſlature.

I always thought it my Duty to join my little Endeavours, to theirs for the ſame Purpoſe; for which Reaſon, in the Choice of every Kind of Materials employ'd in *Weſtminſter Bridge*, I had as great a Regard to *Cheapneſs* as was conſiſtent with *Propriety*, *Solidity*, *and Durableneſs*, all which ſhould be well conſidered in the projecting of all Buildings of this Sort, and in which I had the impartial Advice and Aſſiſtance of ſome of the moſt skillful and moſt experienc'd Builders, and Maſter-Artificers in their reſpective Branches. Two or three Inſtances of this, will, I dare ſay, be read with Pleaſure.

All the ſecondary Arches over thoſe of *Portland Stone*, and the Spandrels of all the Arches in this Bridge, have been built with *Purbeck*

Stone

Stone, for theſe Reaſons, 1*ſt*. Becauſe that Stone is not only as fit, but even fitter for that Purpoſe than *Portland Stone*, being as durable and harder. 2*dly*. Becauſe its Colour being greyer or darker, the Spandrels make a graceful Back-ground, to ſet off the better the Whiteneſs of the *Portland*, in the Turrets, Arches, and Superſtructure. And laſtly, becauſe it is hardly more than half the Price of *Portland*, when worked up and ſet. The Savings obtained thereby on theſe Articles may eaſily be proved to amount to 10,500 *l*. and upwards.

In the Timber made Choice of for all the *Piles*, *Caiſſons*, *Gratings*, *Centers*, &c. no *Oak Timber* was uſed, but where it was abſolutely neceſſary; in every thingelſe ſome kind or other of *Firr Timber* was made Uſe of. 1*ſt*. Becauſe *Firr Timber* is as fit for all the Purpoſes it was employed in as any other. 2*dly*. Becauſe it is as durable, when always kept wet. 3*dly*. Becauſe it is eaſier to be come at; and laſtly, Becauſe the Expence of the firſt Coſt, and Workmanſhip taken together, was thereby leſſened full one Half of what it muſt have been, if *Oak Timber* had been every where made uſe of, as ſome Perſons propoſed: The Savings which this Choice produced, (as may ſoon be made out) have amounted to 7000 *l*. and upwards.

Even as to the Number of Centers that have been uſed to turn all the Arches, I think proper to inform the Readers of a Particular, in which (as in a great many others) the Con-

ſtruction

ſtruction of *Weſtminſter Bridge* has been different from that of other Bridges : In order to which it muſt be obſerved, that it has been the conſtant Uſage in the Conſtruction of Bridges built with Stone or Brick, that conſiſt of many Arches, to frame and ſet up a wooden Frame called a *Center*, under every Arch that is to be turned and built over it, and to eaſe and ſtrike none of them till they are all built, or at leaſt keyed : and both the Abutments of the Bridge compleated and ſettled; becauſe thoſe Arches being generally built the common Way, that is, equally thick in every Part, every one of thoſe Arches, (let their Figure be what it will) and even the Semicircular, do exert a conſiderable *Thruſt*, or lateral Preſſure, againſt their Piers, and againſt one another ; ſo that the Safety of thoſe Bridges greatly depends on the Goodneſs, Largeneſs, Solidity, and Reſiſtance of the Abutments. But as I did not think it proper to rely ſo much on the Goodneſs of the Abutments for many Reaſons, and had therefore contrived the Arches of *Weſtminſter Bridge* (by means of the ſecondary Arches over thoſe of *Portland*, and the ſuperincumbent Weight properly diſpoſed) in ſuch a Manner, that their *Thruſt*, or lateral Preſſure, was thereby counterballanced and deſtroyed. This furniſhed me with another Means of ſaving both Time and Money in the Article of Centers, which I kept to myſelf till a proper Occaſion offered, which was this. As ſoon as the Commiſſioners had laid aſide their firſt

Scheme

Scheme of a wooden Superſtructure upon Stone Piers, and a Bridge entirely Stone was reſolved upon; I propoſed to the Board, (as my Opinion, founded on many reaſons not at all material at preſent) that it were beſt to contract with their Maſter Carpenter, for the making and ſetting up the three Middle and largeſt Centers, and with their Maſter Maſons for building thoſe three Arches, before any of the others; which was approved of, and carried into Execution accordingly; I directed that theſe three Arches ſhould ſtand as they had been built upon their Centers, without eaſing or ſtriking them, my obvious Intention therein, being to obtain thereby a ſteady fix'd Point, as it were, and a temporary Abutment, for the greater Security of each Set of Arches, contained on each Side, between the three Middle ones and each Abutment; then I adviſed the Commiſſioners to proceed in building the whole Set of Arches wanting on the *Weſtminſter* Side, before any of the oppoſite Set ſhould be taken in Hand, which being approved, five more large new Centers, and a ſmall one for the Abutment Arch were contracted for, made, and ſet up, and the Maſons likewiſe built, according to Contract, all thoſe Arches as faſt as the Centers were ſet up. *Then and not before*, I openly declared to the Board, that no more Centers were neceſſary; that the two Centers under the Weſtern 68 and 64 Feet Arches might, and ought to be eaſed, and ſtruck, and then intirely taken down,

down, and ſet up again on the other Side, for the Building of the correſponding Eaſtern Arches, and that the Arches being ſo contrived, as not to exert any conſiderable *Thruſt*, or lateral Preſſure, all the other Centers (excepting the three Middle ones) might be made to ſerve the ſame Purpoſes over again on the Eaſtern Part of the Bridge. In this I was for ſome Time violently oppoſed from a very obvious Quarter, and for very obvious Reaſons; but being convinced that I was in the right, I reſolved to be ſteady, and gave my Directions accordingly; which being in ſome Meaſure demurred to, the Matter was brought before the Board, where the only Reaſon given for having a Center under every Arch, was to plead Cuſtom, and the Danger that might (perhaps) enſue by departing from it. The Board ſeeing thro' all this, enforced the Directions I had given, by a poſitive Order of theirs; the Centers were removed, and every thing anſwered perfectly well, by which I had a triple Satisfaction. The Conſciouſneſs of having done my Duty, the ſaving to the Publick above 5000 *l.* in this ſingle Article of Centers, beſides Time; and thereby pleaſing the Commiſſioners who employed me, and every body elſe, except a few Perſons, who (did not loſe, but only) miſſed getting a good round Sum.

Before I mention what *Weſtminſter Bridge* (as I am informed) has really coſt, I will mention what ſome very ſenſible People were of Opinion

nion it would coſt. At one of the Meetings of the Gentlemen of *Weſtminſter*, who made Application to Parliament for having a Bridge, I heard a Nobleman, whom I dont chuſe to name, (tho' he has been dead many Years) ſay, that from the Knowledge he had in Buildings, from what he had ſeen perform'd in the ſame Way abroad, and from the very great Difficulties and Expences, that he foreſaw would attend the Building of *Weſtminſter Bridge* in Stone, he was clearly of Opinion, that ſuch a Bridge would be at leaſt 20 Years in building, and would coſt at leaſt 400,000 *l.* and perhaps 500,000 *l.* To this Aſſertion, one of the Gentlemen, who ſpoke on the other Side, made no other Anſwer than to ſay, that his Lordſhip might be miſtaken, and if not, he was for his Part, as clearly of Opinion, that even ſuppoſing the Conſtruction of *Weſtminſter Bridge*, in Stone, ſhould require 20 Years, and ſhould coſt 500,000 *l.* it was every way preferable, and much more ſuitable to the Place, than any wooden Superſtructure whatſoever.

Another Nobleman, (deceaſed not long ago, to my very great Sorrow) well known for his Skill in Architecture, has often declared it as his Opinion, that ever ſince it was reſolved that *Weſtminſter Bridge* ſhould be all Stone, he never believed that it would, or could be made paſſable in leſs than 12 or 14 Years; and that if the Expence thereof amounted to 300 000 *l.* he ſhould

ſhould not think it an unreaſonable Price for ſuch a Purchaſe.

Now I am (I have good Reaſon to think well) inform'd, that the total Amount of all the Bills and Contracts, for Materials delivered, Work done, and Labour of all Sorts, in and about *Weſtminſter Bridge*, does not exceed 218,800 *l.*

And that this Bridge was made not only paſſable, or ſerviceable, but compleatly finiſhed, in leſs than 12 Years.

Laſtly, ſince the beſt Means to examine whether any Fabrick is cheap or dear in its Kind, or whether it has been ſlowly or expedidouſly erected, is by comparing it with others, I ſhall take the Liberty, before I conclude, to compare the Money and Time, employ'd in building *Weſtminſter Bridge*, with the Money and Time employed in building the *Cathedral* of St. *Paul's*, in *London*; not only becauſe each of theſe Fabricks, (tho' as different every Way in their Kinds, as in the Uſes that are made of them) hath taken many Years in Building, and coſt great Sums, but becauſe they have been both finiſhed within the Memory of Man, in the ſame Capital, and conſequently, that the prime Coſt at leaſt, of the Materials, and the Price of Man's Labour employed therein, may be ſuppoſed nearly the ſame.

The Facts relating to St. *Paul's*, are, That it was built on dry Land, without any Tide or other Obſtruction, in laying its Foundations, where

where they could employ, and ſet as many Materials, as could be procured; and I make no Doubt, but in that Particular, all was done that could be done, under ſo learned and experienc'd an Architect and Surveyor, as was the late Sir *Chriſtopher Wren*, for whoſe Memory, I join with every Perſon that knows any thing of him, in entertaining the higheſt and moſt ſincere Veneration.

That the Time employed in the Building that moſt beautiful, as well as moſt judiciouſly contrived Fabrick, was upwards of 35 Years, the firſt Stone being laid by Sir *Chriſtopher Wren*, in 1675; and the laſt by his Son *Chriſtopher*, in 1710, by his Father's Direction; as I find it mentioned in the *Parentalia*, or *Memoirs of the Family of the Wrens*, publiſhed by his Grandſon, Mr. *Stephen Wren*.

And that the Total neat Expence, as may be ſeen in *Maitland*'s *Hiſtory of London*, did not exceed 736,800 *l*. Now if it be conſidered, that tho' the Quantity of Stone uſed in St. *Paul*'s, is not above half what has been uſed in *Weſtminſter Bridge*, yet as there has been ſo much Workmanſhip, and ſo exquiſite in every kind, beſtowed on a great Part of the Materials at St. *Paul*'s, I am almoſt ſure, that every Perſon properly qualified to judge of theſe Things, will join with me in Opinion, that upon the whole, St. *Paul*'s *Cathedral*, is not only one of the moſt magnificent, but one of the cheapeſt, and one of the moſt expeditiouſly built in *Chriſtendom.*

The only Facts relating to *Weſtminſter Bridge*, which are neceſſary to be known, in order to make the Compariſon, (as to Time and Expence) are theſe :

That *Weſtminſter Bridge* contains near twice the Quantity of Stone Materials, contain'd in St. *Paul*'s ; that its Foundations are all laid in Water, ſeveral Feet below the Bed of the *River Thames* ; where, till all the Piers and Abutments were brought up above High-water Mark, the Tides twice in 24 Hours, did and muſt neceſſarily ſtop or retard the Progreſs of the Building.

That the whole Fabrick was begun and finiſhed in leſs than twelve Years.

And that the whole Amount of the neat Expence that has attended the Building of *Weſtminſter Bridge*, does not exceed 218,800 *l.*

I will not foreſtall the Readers in the Pleaſure of pronouncing the Reſult of the Compariſon, and conclude with expreſſing my ſincereſt Wiſhes for the Health and Proſperity of all the good People, for whoſe Service and Conveniency *Weſtminſter Bridge* has been built.

The E N D

THE

PLAN

OF A

WORK

Intended to be PUBLISHED by the AUTHOR,

BUT NOT BY SUBSCRIPTION,

As ſoon as finiſhed.

A

DESCRIPTION

OF

Weſtminſter Bridge.

To which are added,

HISTORICAL ACCOUNTS

RELATING TO THE

Building and Expence thereof.

WITH

TECHNICAL DESCRIPTIONS of all the OPERATIONS, MACHINES, ENGINES, &c. made uſe of in the Courſe of the WORKS.

ALSO

Analitical Inveſtigations, Calculations, and Geometrical Conſtructions, referr'd to in the former Parts; with Practical Rules, and Obſervations in Mechanics, Hydraulics, and the Art of Building in Water.

Illuſtrated with a great Number of COPPER-PLATES.

Bid Harbours open, public Ways extend;
Bid Temples, worthier of the God, aſcend;
Bid the Broad Arch the dangerous Flood contain,
The Mole projected break the roaring Main,
Back to his Bounds, their ſubject Sea command,
And roll obedient Rivers thro' the Land. POPE.

LONDON:
Printed for the AUTHOR.

TO THE

READER.

THE Two principal Motives which induced me to undertake this Work are:

First; The repeated Inſtances of ſeveral eminent Perſons; many of them no leſs conſpicuous for their Birth, Rank, and Fortune, than for their Learning, and other valuable Qualifications: All whoſe Requeſts it will ever be my Ambition to comply with; and eſpecially, to give them this publick Teſtimony of my Gratitude, and of the Veneration which I entertain for them.

Secondly: As it is the Duty of every Man to do all the Good in his Power, and to contribute, ſo far as his Abilities will permit, to the Welfare of Poſterity; I thought it incumbent on me to communicate to Others ſome of the Fruits of my Studies, join'd to above twenty Years Practice and Obſervation.

It is not more generally acknowledged than deplor'd, that we have ſcarce any Account extant of the Methods obſerved in the Conſtruction of Bridges, and other conſiderable publick Works. And it is as certain that Princes and States, as well as private Perſons, can never be able (without ſome Documents or Inſtructions, like to thoſe here intended) to form any Judgment with regard to the Succeſs, to the Practicableneſs, or to the Time and Expence which the Execution of ſuch Works muſt naturally require.

Let me add; that as the Perſons more immediately concern'd in *Deſigning*, Directing, Conducting, and Executing ſuch Works, have ſo few and ſuch imperfect Accounts of what has been already perform'd in the like Caſes, and of what Methods have proved ſucceſsful (or otherwiſe) in the Execution; they thence are often obliged either to uſe Expedients, of the Succeſs whereof (for want of ſufficient Trials) they cannot have any Certainty; or elſe muſt be forced to employ ſuch old Methods as they themſelves are acquainted with, tho' often leſs proper, and attended with a greater Expence of Money and Time.

In order to render this Work as commodious with regard to Size, and of as eaſy Purchaſe as poſſible, it will be divided into two Volumes Octavo, excluſive of the Plates, each Volume to conſiſt of two Parts.

A Table of their Contents is here annex'd, whereby the Reader will easily perceive, that the first is drawn up (purposely) so as to contain a compleat Description of *Westminster Bridge*, in all its Parts; with the History of the Building, and Expence thereof. This first Volume will (I humbly conceive) be sufficient for the Generality of Readers.

The Second Volume is intended chiefly for the Use of Engineers, Architects, ingenious Artificers, and such other Persons, who are desirous of being acquainted with the Principles, on which the Works of *Westminster Bridge* are founded; the Methods whereby they were executed; and the Application of Mechanicks, Hydrostaticks, and other Branches of pure and mixed Mathematicks to Practice.

As the Number of Plates proper to illustrate the Subject-Matter of each Volume will be very considerable, and the Size of those Plates must necessarily be very different, in order to render the Objects represented by them sufficiently plain; I am of Opinion, that the Plates belonging to each Volume ought to be kept separate; and not folded, bound up, or sew'd in with the Books, which will greatly contribute to their Preservation. I thought this Caution necessary; since, as several of the Plates are Perspective Views of the whole Bridge, of some of its principal Parts, or of the chief Machines used in the Building, they will not only be instructive, but likewise ornamental.

As particular Care will be taken, and no Expence ſpared, to procure fine Paper and the neateſt Types, and to have the Plates very correct, and engraved with the utmoſt Neatneſs and Beauty; the Deſcriptions will be brought into a much ſhorter Compaſs than could otherwiſe be done conſiſtently with Perſpicuity; ſince the Reader will, by Means of Scales inſerted in every Page, be able to meaſure the true Dimenſions of any Part which may not be particularly deſcribed.

The original Draughts of all the Plans, Elevations and Sections; ſeveral Models; all the Eſtimates and Calculations; moſt of the neceſſary Vouchers; and a regular Journal of the Progreſs of the Works, free from all trifling Particulars, are my own, and ſhall remain in my Cuſtody, till this Work is compleated.

But in Caſe I ſhould be prevented in this Deſign by Death, I have bequeathed, by my Will, all that I ſhall die poſſeſſed of (relating in any Manner to the future Preſervation of *Weſtminſter Bridge*, or to the Hiſtory of its Conſtruction) to Perſons, to whoſe noble Spirit and Patience the Publick in general, and the Commiſſioners in particular, muſt acknowledge themſelves obliged. for having not only *a Bridge at Weſtminſter*; but alſo for having it (perhaps) the beſt built, and certainly one of the moſt magnificent Bridges in *Europe*. In doing this I have had the ſingular Pleaſure of perfectly reconciling

ciling my own ſtrong Inclination with Duty and Gratitude.

My Intention, in Printing and diſtributing this Plan among my Friends and Acquaintance, is, in the firſt Place, to eaſe the Minds of many Perſons who ſeem to fear, that ſufficient Care has not been taken to collect and preſerve the proper Materials for a full Deſcription of this Bridge, with the Hiſtory of its Building; and effectually ſecure to Poſterity, (by Models and original Drawings) the many new and uſeful Inventions which have been made uſe of in its Execution.

Laſtly: Since *Weſtminſter Bridge* has been (ſome few Ornaments excepted) intirely built upon my own *Deſigns*, according to Methods of my own contriving; I think proper to forewarn People not to lend their Names; or ſuffer themſelves to be taken in for any *Subſcription-Money*, by Perſons perhaps no ways qualified for ſuch a Work as this: The Right Honourable, &c. the Commiſſioners, and Others, well knowing that I have an *indiſputable Right* to expect that, after the Bridge is finiſhed, I ſhall have Leiſure and Means ſufficient, to prevent (as I fully intend it) Their being at any Expence on this Account.

Charles Labelye.

Weſtminſter,
July 29, 1744.

P. S. Since the first Writing and Pulishing this Preface, and the following Plan of a Book, which was in *July* 1744, God Almighty having been pleased to take unto himself several Noblemen and Gentlemen, who had all shewn an uncommon Publick Spirit and Patience, in the promoting, forwarding, and well conducting the Building of *Westminster Bridge*; I have been obliged to alter my Will, as to the Persons to whom I have bequeathed whatever I should happen to die possest of, (relating in any Manner to *Westminster Bridge*) in Case I should be prevented by Death, from compleating this Work.——And as the Right Honourable *&c.* the Commissioners, according to their Promises, and without any Solicitations from me, have been pleased to afford me both *Leisure and Means sufficient*, to enable me to go on with it; I conclude by assuring them and the rest of my Readers, that I shall proceed therein without Delay, as well, and as fast as my Health (which has been declining for some Years) will permit.

Charles Labelye.

Westminster,
April 30, 1751.

TABLE of the CONTENTS OF VOLUME I. PART I.

Including a general DESCRIPTION of

WESTMINSTER BRIDGE,

And particular DESCRIPTIONS of its

SEVERAL PARTS,

Under the following HEADS or CHAPTERS.

VOL.

VOLUME I. PART II.

Containing ſeveral Hiſtorical Accounts relating to the Building of WESTMINSTER BRIDGE, *with the Expence thereof.*

Under the following HEADS and CHAPTERS.

VII. In what Manner the Accounts were examined, and audited by the Right Honourable, &c. the Commiſſioners; and laid yearly (together with Copies of Contracts, and all other material Tranſactions) before both Houſes of Parliament.

VIII. A ſuccinct Account of the chief Tranſactions relating to *Weſtminſter Bridge*, before the Right Honourable, &c. the Commiſſioners had fixed upon any *Deſign*.

IX. The moſt material Tranſactions and Occurrences in the Courſe of the Building, extracted chiefly from the AUTHOR's Journal.

X. An exact Account of the Expence of *Weſtminſter Bridge*, diſtinguiſhed under proper Heads, and compared with the AUTHOR's Eſtimates.

XI. Remarks upon *Weſtminſter Bridge*; and Juſtice done (ſo far as lies in the AUTHOR's Power) to the ſeveral Perſons concerned in its Execution.

APPENDIX.

A ſhort Account of the new Streets, Roads, &c. leading to, and from *Weſtminſter Bridge*; and the Expences thereof diſtinguiſhed under proper Heads.

VOLUME II. PART I.

Containing Technical Descriptions of the ſeveral Operations, Machines, &c. *made uſe of in building* WESTMINSTER BRIDGE.

Under the following HEADS or CHAPTERS.

I. MEaſuring the Breadth of the *River Thames*, by actual Menſurations.

II. Sounding the River, with an Account of the Soundings above, at, and below *Weſtminſter Bridge*.

III. Setting out the Encloſures of Fenders, or Guard-Piles, to ſecure the Building of the Piers.

IV. The Engine for Driving the Piles, and the Manner of Driving them deſcribed.

V. The Manner of Boring under the Bed of the River, with an Account of all the Borings under the Piers and Abutments.

VI. The Methods obſerved in Digging, Levelling, examining and preſerving Level, the Foundations of the Piers under Water.

VII.

VII. The Conſtruction of the *Cheſts* or *Caiſſons*, The Manner of Launching them off the Stocks, and the Methods of placing their Sides on new Bottoms; with ſeveral other Ways of moving on dry Land, or ſetting on float ſuch unwieldy Bodies at a ſmall Expence.

VIII. The Methods obſerved in the Building of the Piers, and ſetting them down in their Places, from the Bringing the *Cheſts* or *Caiſſons* within the Encloſures of Guard-Piles, to the Lifting of their Sides over the ſaid Piers when built, and all the Operations deſcribed in their proper Order. This Chapter concludes with Hints relating to ſeveral conſiderable Works, to which the Methods of building in *Cheſts* or *Caiſſons* may be uſefully applied.

IX. A Deſcription of the ſeveral Methods made uſe of in laying the Foundations of the Abutments.

X. The Centers for turning the Arches; and the ſeveral Ways made uſe of for Eaſing and Striking them deſcribed.

XI. The Methods obſerved in building the Arches and Superſtructure, with an Account of the Nature, Quality, and Diſpoſition of the ſeveral Materials employed therein.

XII. Descriptions of the several Ways made use of in drawing up the Piles.

XIII. The Machines for Sawing of the Piles under Water, and the Manner of Sawing them off described.

XIV. A particular Account of the Settling of one of the Piers of *Westminster Bridge*. The most likely Causes of this Accident.

XV. Some Accounts of the several Schemes proposed for repairing the damaged Pier, and the two adjoining Arches: And lastly, an Account of the Methods which were proposed by the Author for the same Purpose, and have proved successful in the Execution.

VOL.

VOLUME II. PART II.

Containing the Analytical Investigations, Calculations, and Geometrical Constructions referred to in the former Parts; with practical Rules and Observations in Mechanicks, Hydraulicks, and the Art of building in Water.

Under the following HEADS or CHAPTERS.

IV. Obſervations and Calculations relating to the Engine for driving Piles; ſhewing what it has in common with other Driving Engines, wherein it is preferable to moſt, and how it may be made much more ſimple and leſs Expenſive.

V. Several Calculations relating to the *Cheſts* or *Caiſſons*, ſuch as the Quantity of Timber and other Materials, The Weight of an intire *Cheſt* or *Caiſſon*: It's Draught of Water when empty: The Contents of it's Inſide: The Total Preſſure of the ambient Water, when the Sides are fix'd to the Bottom, and the moſt advantageous Places for ſtrutting the Sides againſt that Preſſure determined.

VI. Obſervations relating to the lifting the Sides of a *Cheſt* or *Caiſſon* over the Piers when built, and the Force of the Tide in all ſuch Caſes computed.

VII. The Quantity of the ſeveral Materials contained in each of the Piers, Arches, and Abutments of this Bridge, the Weight of each calculated, and the whole ſummed up.

VIII. Calculations relating to circular and ſtreight Wedges, with Remarks on the Uſe of thoſe ſeveral Ways.

IX.

IX. General Obſervations upon Arches. Vulgar Errors on that Subject expoſed. The overſights of ſeveral Mathematicians and others, in their Application of certain Properties of the Curve, called the *Catenaria* to Arches. Reaſons why the Problem for determining the *Thruſt* or lateral Preſſure of all Arches in general, has hitherto been attempted in Vain. The Miſtakes of *Palladio*, *Blondel*, *Gaultier*, and all thoſe who have already given, or ſhall ever pretend to give general Rules on that Subject. The *Thruſt* or lateral Preſſure of various Kinds of Arches (given in all their Dimenſions) determined, by Calculations deduced from ſuch Suppoſitions as do really occur in Practice. Reaſons why no general Rule can be drawn from the Solutions of ever ſo many particular Caſes. Several of thoſe pretended general Rules proved to be not only abſurd, but extremely dangerous in the Execution. The Application of this Doctrine to the Arches of *Weſtminſter Bridge*. The Curve and different Thickneſſes of its ſecundary Arches (by means of which, together with the ſuperincumbent Weight properly placed, the *Thruſt* or lateral Preſſure of thoſe Arches is intirely counterballanced and deſtroyed) calculated and conſtructed, with ſeveral Obſervations.

X. Calculations and Obſervations relating to the Machines employed in hoiſting up the ſeveral Materials at *Weſtminſter Bridge*, compared with the *Cranes* commonly uſed in *England*, and with the *French Grues*. The *Maximum*, or the greateſt Effect that can be expected from any Machine worked by Men or Horſes. Vulgar Errors on this Subject expoſed, together with the Miſchiefs often ariſing from them.

XI. Obſervations, and practical Rules relating to the Subject of raiſing Water, and draining; with Remarks on Chain Pumps, and moſt other Pumps.

XII. Calculations relating to the ſeveral Ways of drawing up the Piles, and the neceſſary Force computed.

XIII. Obſervations and Calculations concerning the Sawing of Piles under Water.

XIV. Rules forCalculating the Fall occaſioned in a River or Stream of Water, by the Piers of a Bridge, exemplified in the Falls under the Arches of *Weſtminſter* and *London Bridges*, and the Reſults of the Calculations, compared with the Obſervations made on thoſe Falls.

XV. Remarks on the various Motions and Actions of the Streams and Tides, particultarly in Rivers that have flat Shores, Counter Tides, and Eddies, ſuch as the *River Thames* above and below *Weſtminſter Bridge.*

The End of the Second and Laſt Volume.

[illegible] Memorandum [illegible] and [illegible] of the [illegible] Tides, [illegible] in Rivers that have their Shores Composed of Docks and [illegible] as the River Thames above and below Westminster Bridge.

The End of the Second and Last Volume.

Printed in the United States
By Bookmasters